Complete Guide to Amateur Radio

Complete Guide to Amateur Radio

Joseph DuBovy

PARKER PUBLISHING COMPANY, INC.
WEST NYACK, N.Y.

©1979, by

PARKER PUBLISHING COMPANY, INC.
West Nyack, N.Y.

All rights reserved. No part of this book may be reproduced in any form or by any means, without permission in writing from the publisher.

Library of Congress Cataloging in Publication Data

DuBovy, Joseph.
   Complete guide to amateur radio.

   Includes index.
   1. Amateur radio stations.   2. Radio—Amateurs' manuals. I. Title.
TK9956.D76   621.3841'66   79-13374
ISBN 0-13-159921-6

Printed in the United States of America

DEDICATION

Dedicated to Barbara DuBovy, whose interest in Amateur Radio inspired the original research that made this book possible.

2070740

ACKNOWLEDGEMENT

I would like to acknowledge the many people at the American Radio Relay League for their generous assistance and helpful advice.

A Word from the Author on How This Book Will Help You

There are many ways to illustrate the practical advantages Amateur Radio has over CB and other forms of radio. For example, over the past couple of decades many people have become acquainted with CB, and operators now have more channels to work with. The sets are limited by law to a maximum power of 5 watts, and there is cause for celebration when they reach out beyond 15 or 20 miles to deliver or receive a message.

Compare this with that incredible moment when you're able to throw the switch on your Amateur Radio and have clear, enjoyable conversations with a colleague in London because you're operating a transmitter with a power input of up to 2,000 watts and a dial spread of approximately 100,000 "channels," compared with the 40 available on CB. Truly, there is a magic and power in Amateur Radio that makes this one of the most rewarding, fascinating subjects in the entire field of electronics. The anticipation of talking with someone in Europe over your set, coupled with the essential guidelines you'll find in this book, will provide you with the guidance you need to pass all classes of Amateur Radio FCC exams. Here, indeed, is a *complete*, practical guide to the remarkable advantages offered by this form of "long distance" radio.

In Chapter 1 you will acquire a clearer understanding of the factors that make possible this branch of radio. You'll learn about spark gaps, crude oatmeal box receivers, AM, FM, and SSB (Single Sideband) modulation, and about ham wavelengths that are longer than some AM broadcast stations, and shorter than radar. You will read about how you can set up your own station to send teletype to the far corners of the world. In this chapter, the key fundamentals of ham television will be explained. If you wish, you will eventually be able to transmit slow-scan video to other hams with similar equipment. In Chapter 1 the various classes of FCC licenses, their requirements, and their privileges will be presented in detail.

In Chapter 2 you will learn more about the peculiar propagation characteristics for various frequencies, such as how it is possible to talk to the Far East with low power at certain times, when atmospheric conditions are just right. Also why, despite high power, you cannot talk to far distant points when atmospheric conditions do not favor your signal.

In Chapter 3 you will learn how amateur radio enthusiasts, via the American Radio Relay League, helped to make possible a government missile to launch OSCAR I (Orbiting Satellite Carrying Amateur Radio), and how Oscar 8 passes 560 miles overhead twice daily with its broadband receiver waiting to relay *your* voice to distant parts of the earth.

A clear, basic overview of both simple and complex transmitter and receiver circuits in Chapter 4 will enable you to quickly grasp the essentials of even the most sophisticated modern communications equipment.

Starting with the basic doublet, Chapter 5, in simple terms, explains the principles and mechanics of the most frequently used ham antennas. Whether you are interested in a large antenna array or a tiny transmitting antenna for your small apartment, it will be described in this chapter.

Later, in Chapter 12, you will be able to put Chapter 5 to practical use. Construction details are described that will enable you, even if you don't have a technical background, to build one of the most efficient amateur antennas in use today. Among these antenna projects is the "satellite signal-squirter." This antenna, made of one foot long welding rods, weighs less than half a pound. Held in your hand, it will permit you to trigger amateur satellites with a walkie-talkie. Imagine talking to a far away country while standing on a street corner!

With all of these exciting possibilities available to the ham, how difficult will it be for you to get your license? In Chapter 9, you are given typical questions found on the Novice Class FCC exam. This material should be viewed as a reliable guide to further study so that the theory intrinsic to each question is fully understood.

A WORD FROM THE AUTHOR

CB operators by the tens of thousands sometimes find it difficult to muster the courage to take the FCC amateur radio exam because of their fear of Morse code. During World War II, the U.S. Navy found itself in a critical situation. In order to keep the fleet sailing, the Navy had to make raw recruits into skilled telegraph operators in a short time. The Navy called upon leading psychologists to devise a code course that would solve this serious problem. In fact, the course that was subsequently designed enabled thousands of men who had never seen a telegraph key before to operate a ship's telegraph transmitter in *two weeks time*. Chapter 10 is a detailed study guide to the Navy's famous course. In addition, it offers instructions on how to build your own code practice oscillator and code practice buzzer. Also listed are ways to find the short wave code practice transmissions that will allow you to evaluate your own progress.

After passing the Novice exam, you will be able to proceed to the Technician and General class license tests so that you can use *all* the ham bands. Chapter 13 contains questions that are used in these advanced exams. If you read Chapters 6, 7, and 8 diligently, you should have no difficulty in understanding the typical FCC exam questions and answers listed in Chapter 13. Using "Oscar" requires some patience, as this satellite is in constant motion at 15,000 miles per hour. One alternative is to trigger one of the 4,000 mountain-top repeater stations with your walkie-talkie or automobile rig. Repeaters are installed by local radio clubs and can also be located on high skyscrapers. They are capable of transforming a one watt walkie-talkie into a powerful radio broadcast station. Chapter 14 tells you how to use repeaters.

After becoming an amateur with all the advanced operating privileges, you will want to maintain your equipment at the highest operating efficiency. The last Chapter describes simple test equipment you can either build or buy that will allow you to get the most out of your ham gear.

With this book, you will be able to capitalize on the marvelous opportunities available to all those who master the challenges offered by amateur radio. I'll be listening for you on the air.

Joseph DuBovy, W2TCC

Contents

    A Word from the Author on How This Book Will Help You.. 7

1   The Exciting World of Ham Radio 19

    The Pioneers • 19
    Amateur Licenses • 20
    Specialized Communication • 31
    Amateur Television (TV) • 31
    Slow-Scan Television (SSTV) • 32
    Phone Patching • 34
    Satellites • 34
    Radioteletype (RTTY) • 36
    Frequency • 38
    Wavelength • 39
    Meters and Megacycles Over the Radio
       Frequency Spectrum • 40

2   Understanding the Ham Bands and Propagation ... 43

    Radiation from the Sun • 43
    Sunspots • 44
    The Ionosphere • 45
    The Layers of Ionization • 45
    Propagation Bulletins • 49
    The Low Frequencies • 49
    The High Frequencies • 50

3   Talking to Europe on Your Walkie-Talkie 53

    Space Communication for Everyone • 53
    The History of Amateur Satellites • 53
    Equipment Required to Talk to OSCAR • 54
    Europe on a Walkie-Talkie • 58
    Space Experiments • 58
    Locating OSCAR • 60

The Phase III Series • 62
The Technical Problem • 64
Amateurs and Professionals • 64
The Mechanics of Phase III • 65
Best Frequencies • 66
The Power Supply • 67
Continuing Control via Microprocessors • 68
The Amateur/Professional • 70
The Cost of Satellites • 70
OSCAR in the Classroom • 71
Other Satellite Applications • 72

4   A Practical Guide to Basic Ham Circuits: How They Work 75

How the Crystal Set Became the TRF Receiver • 75
The Super-Het • 78
FM Receivers • 80
Receiving Morse Code and Single Sideband • 83
Single Sideband • 83
Transmitters • 84
The Spark Gap Transmitter • 84
FM Transmitters • 87
Single Sideband Transmitters • 88

5   How to Understand Antennas 91

Basic Theory • 91
Antenna Current • 91
Distribution of Voltage and Current • 93
Impedance • 94
Feeding Energy to the Antenna • 95
Vertical Antennas • 97
Loading Coils • 98
Polarization • 100
Directivity • 100
The Yagi • 103
Helical Whip • 104
The Cubical Quad • 106
Inverted V • 106
Collinear and Parabolic Arrays • 109

CONTENTS

Masts • 110
Transmatch • 111

**6 A Practical Guide to Schematic Diagrams
    and Basic Electron Theory 113**

Electron Theory • 115

**7 Working with Semiconductors and Other
    Components 133**

Semiconductor Theory • 133
Junctions • 133
Diodes • 134
Transistors • 136
Quartz Crystals • 137
Field Effect Transistors • 138
Resistors • 139
Capacitors • 139
Chokes • 141
RF Inductors • 142

8 Simplified Guide to Amateur Circuits in Detail .. 147

High-Pass, Low-Pass, and Band-Pass Filters • 147
High-Pass Filters • 147
Low-Pass Filters • 148
Band-Pass Filters • 149
Power Supplies • 150
Surge Protection • 153
Oscillators • 154
Synthesizers • 156
RF Amplifiers • 158
RF Tanks and Antenna Coupling • 161
Pi-Output Tanks • 165

9 Typical Questions on the Novice FCC Exam 167

Rules and Regulations • 167
Radio Phenomena • 170
Operating Procedure • 171
Theory of Electricity • 173

Circuits • 174
Electronic Components • 176
Are You Ready for the Exam? • 180

10 How to Benefit from the Amazing Navy Code Study Course 183

Why Code? • 183
The Navy's Method • 184
A Morse Code Oscillator • 186
Rules for Code Practice • 187
Rules for Sending • 188
Holding the Key • 189
Adjusting Your Key • 190
Warming Up to the Key • 191
Exercises • 191
Knowing When You Are Ready • 192

11 How to Build Your Own Set 195

In the Early Days • 195
A Homemade Transmitter • 195
Construction Practice • 196
Soldering • 200
The Triode-Pentode • 200
The Project • 201
The Power Supply • 202
Tuning Up • 203

12 A Guide to Great Antennas You Can Easily Build 205

The 80 Meter Doublet • 206
Wire Size • 208
The Balun • 208
Testing the Antenna • 209
The Short 40 Meter Vertical • 210
Loading Coil • 211
A Satellite Signal-Squirter • 215
Two-Meter Antennas • 217

CONTENTS

13 Preparing for the General and Technician's Exam **219**

Technician's License • 219
General Class Exam • 219
Sample Questions • 220
Rules and Regulations • 220
Propagation and Operation • 223
Basic Electricity • 227

14 Repeaters and How to Understand Them **239**

Finding Repeaters • 239
The Machine • 239
Remote Base Stations • 242

15 Key Factors in Amateur Test Equipment **245**

Introduction • 245
Electron Voltmeters • 245
RF Probe • 246
The Oscilloscope • 247
Frequency Marker • 250
Audio Oscillators • 252
Field Strength Meter • 254
Measuring VSWR • 255
Noise Generators • 257

Index **259**

1

The Exciting World of Ham Radio

THE PIONEERS

At the turn of the century there were a few amateur operators who experimented with crude spark gap transmitters sending signals over short distances. In 1904 the galena crystal detector was invented and a short time later Marconi sent a telegraph signal across the Atlantic. This feat created a great deal of publicity, and the popularity of amateur radio increased sharply. Today there are some 300,000 amateurs in the United States, and almost a million throughout the world. Amateurs are self-trained experimenters who pursue their hobby without financial interest.

Amateurs have won the gratitude of the nation for heroic performances in times of natural disaster. Since 1913 amateur radio has been the principal means of outside communication in storms, floods, and earthquakes in the U.S.A. During tornadoes, forest fires, and blizzards amateurs played the major role in relief work and earned wide commendation for their resourcefulness in effecting communication where all other means had failed.

The American Radio Relay League is the national amateur organization that promotes amateur radio, participates in international radio conferences, and provides vast educational services to amateurs. In 1938 the ARRL organized an emergency corps which cooperates closely with the American Red Cross. In 1961 through Project Oscar, an ARRL affiliate, the first amateur satellite was launched by NASA. The ARRL is the spokesman for U.S. and Canadian amateurs and is owned by all its members. Its monthly publication is called QST. The ARRL maintains a model amateur station, WIAW, which transmits bulletins of general interest to amateurs. WIAW conducts code practice transmissions to help you prepare for your FCC license examination.

AMATEUR LICENSES

Amateur licenses are issued to persons of any age. The license fee is $9.00. When you have studied the theory of Chapter 9 and practiced code according to Chapter 10, you will be ready to take your novice FCC exam at any of the locations listed in Figures 1-1a–1-1c. There are currently four classes of amateur license—Novice, Technician, General, and Amateur Extra Class. The Morse code requirement for both Novice and Technician licenses are the same—five words per minute—while the Morse code requirement for the General and Amateur Extra Class licenses is 13 and 16 words per minute. The written part of the exam is easiest for the Novice test (see Chapter 9) and more difficult for the four, more advanced licenses.

EXAMINATIONS AT FCC FIELD OFFICES

The examination schedule for the FCC field offices is shown below. A prior appointment is necessary:
1. for commercial radiotelegraph examinations where code is required.
2. when specifically indicated in this schedule.
3. for blind applicants.
4. for organized groups of 10 or more.

Otherwise, a prior appointment is not necessary.

Examinations are not conducted on Saturdays, Sundays, or legal holidays.

The FCC field offices are listed alphabetically by state. The field offices in Puerto Rico and the District of Columbia (Washington, D.C.) also are included in the alphabetical list.

EXCITING WORLD OF HAM RADIOS

FCC FIELD OFFICE	EXAMINATION SCHEDULE
ALASKA, Anchorage	Commerical Radiotelephone Examinations Monday through Friday - 8:00 AM to 12:00 Noon Amateur Examinations Requiring code - Monday through Friday by appointment only Not requiring code - Monday through Friday 8:00 AM to 12:00 Noon
CALIFORNIA, Long Beach	Commercial Radiotelephone Examinations Tuesday and Thursday - 8:30 AM through 2:00 PM Amateur Examinations Requiring code - Wednesday - 8:30 AM and 12:30 PM Not requiring code - Wednesday - 8:30 AM through 2:00 PM
CALIFORNIA, San Diego	By appointment only. Appointment should be made one week in advance.
CALIFORNIA, San Francisco	Commercial Radiotelephone Examinations Tuesday and Thursday - 8:30 AM to 2:00 PM Amateur Examinations Requiring code - Wednesday - 8:30 AM and 1:00 PM Not requiring code - Wednesday - 8:30 AM to 2:00 PM
COLORADO, Denver	Commercial Radiotelephone Examinations Tuesday and Thursday - 8:30 AM to 1:00 PM Amateur Examinations Requiring code - Wednesday - 8:30 AM Not requiring code - Tuesday and Thursday 8:30 AM to 1:00 PM
DISTRICT OF COLUMBIA	Commercial Radiotelephone Examinations Tuesday and Thursday - 8:30 AM to 12:30 PM Amateur Examinations Requiring code - Wednesday - 9:00 AM Not requiring code - Wednesday - 9:00 AM to 1:00 PM
FLORIDA, Miami	Commercial Radiotelephone Examinations Tuesday and Wednesday - 8:15 AM to 1:00 PM Amateur Examinations Requiring code - Thursday - 9:00 AM Not requiring code - Tuesday and Wednesday 8:15 AM to 1:00 PM
FLORIDA, Tampa	By appointment only. Appointment should be made one week in advance.
GEORGIA, Atlanta	Commercial Radiotelephone Examinations Tuesday and Friday - 8:30 AM to 12:00 Noon Amateur Examinations Requiring code - Friday - 8:30 AM Not requiring code - Tuesday and Friday 8:30 AM to 12:00 Noon

GEORGIA, Savannah	By appointment only. Appointment should be made one week in advance.
HAWAII, Honolulu	Commercial Radiotelephone Examinations **Tuesday and Thursday - 8:00 AM** **Other times by appointment.** Amateur Examinations **Wednesday - 8:00 AM** **Other times by appointment.**
ILLINOIS, Chicago	Commercial Radiotelephone Examinations **Wednesday and Thursday - 8:45 AM and 1:00 PM** Amateur Examinations **Tuesday and Friday - 8:45 AM**
LOUISIANA, New Orleans	Commercial Radiotelephone Examinations **Tuesday - 10:00 AM to 12:00 Noon** **Wednesday - 8:00 AM to 12:00 Noon** Amateur Examinations **Requiring code - Tuesday - 8:00 AM** **Not requiring code - Tuesday - 10:00 AM to 12:00 Noon** **Wednesday - 8:00 AM to 12:00 Noon**
MARYLAND, Baltimore	Commercial Radiotelephone Examinations **Monday and Friday - 8:30 AM to 12:00 Noon** Amateur Examinations **Requiring code - Monday - 8:30 AM** **Not requiring code - Monday and Friday** **8:30 AM to 12:00 Noon**
MASSACHUSETTS, Boston	Commercial Radiotelephone Examinations **Tuesday and Thursday - 9:00 AM to 11:00 AM** Amateur Examinations **Wednesday - 9:00 AM**
MICHIGAN, Detroit	Commercial Radiotelephone Examinations **Tuesday and Thursday - 9:00 AM to 11:00 AM** Amateur Examinations **Wednesday and Friday - 9:00 AM**
MINNESOTA, St. Paul	Commercial Radiotelephone Examinations **Thursday - 8:45 AM** Amateur Examinations **Friday - 8:45 AM**
MISSOURI, Kansas City	Commercial Radiotelephone Examinations **Wednesday and Thursday - 9:00 AM** Amateur Examinations **Tuesday - 9:00 AM**
NEW YORK, Buffalo	Commercial Radiotelephone Examinations **Thursday - 9:00 AM to 12:00 Noon** Amateur Examinations **Requiring code - Friday at 9:00 AM** **Not requiring code - Friday at 10:00 AM**

EXCITING WORLD OF HAM RADIOS

NEW YORK, New York	Commercial Radiotelephone Examinations Tuesday and Thursday - 9:00 AM to 12:00 Noon Amateur Examinations Wednesday - 9:00 AM
OREGON, Portland	Commercial Radiotelephone Examinations Tuesday and Wednesday - 8:45 AM Amateur Examinations Friday - 8:45 AM
PENNSYLVANIA, Philadelphia	Commercial Radiotelephone Examinations Monday, Tuesday, and Wednesday - 10:00 AM to 12:00 Noon Amateur Examinations Requiring code - Tuesday and Wednesday - 9:00 AM Not requiring code - Monday, Tuesday, and Wednesday 10:00 AM to 12:00 Noon
PUERTO RICO, San Juan	Commercial Radiotelephone Examinations Thursday and Friday - 8:30 AM (or 1:00 PM by appointment only) Amateur Examinations Requiring Code - Friday - 10:00 AM Not requiring code - Thursday and Friday 8:30 AM (or 1:00 PM by appointment only)
TEXAS, Beaumont	By appointment only. Appointment should be made one week in advance.
TEXAS, Dallas	Commercial Radiotelephone Examinations Wednesday and Thursday - 8:00 AM to 11:00 AM Amateur Examinations Tuesday - 9:00 AM
TEXAS, Houston	Commercial Radiotelephone Examinations Tuesday and Thursday - 8:00 AM to 11:30 AM Amateur Examinations Requiring code - Wednesday - 8:00 AM to 8:30 AM Not requiring code - Wednesday - 9:30 AM to 11:30 AM
VIRGINIA, Norfolk	Commercial Radiotelephone Examinations Wednesday and Friday - 9:00 AM to 12:00 Noon Amateur Examinations Requiring code - Thursday - 9:00 AM Not requiring code - Thursday - 10:00 AM
WASHINGTON, Seattle	Commercial Radiotelephone Examinations Tuesday and Wednesday - 8:45 AM Amateur Examinations Friday - 8:30 AM

Figure 1-1a

EXAMINATIONS AT OTHER LOCATIONS

The FCC conducts periodic examinations at a number of cities where it does not maintain offices. It is necessary to make a prior appointment with the FCC field office that gives the examinations. To make this appointment you should:

1. file your application as far in advance as possible with the FCC field office that administers the examination.
2. indicate the city where you wish to be examined.
3. indicate which month you wish to take the examination.

You will then be notified when and where to appear for the examination.

The following list shows these cities, the months during which the examinations are given, and the FCC field offices that administer the examinations. The addresses and telephone numbers of the FCC field offices are shown in Figure 1-1c.

The list is alphabetical according to state. Guam also is included in the alphabetical listing.

STATE	CITY	MONTH IN WHICH EXAMINATION ADMINISTERED	OFFICE ADMINISTERING EXAMINATION
ALABAMA	Birmingham	MAR, SEP	Atlanta, Georgia
	Mobile	JAN, APR, JUL, OCT	New Orleans, Louisiana
	Montgomery	JUN, DEC	Atlanta, Georgia
ALASKA	Fairbanks	JAN, APR, JULY, OCT	Anchorage, Alaska
	Juneau	MAY, NOV	" "
	Ketchikan	MAY, NOV	" "
ARIZONA	Phoenix	JAN, APR, JUL, OCT	Long Beach, California
	Tucson	APR, OCT	" "
ARKANSAS	Little Rock	FEB, MAY, AUG, NOV	New Orleans, Louisiana
CALIFORNIA	Bakersfield	MAY, NOV	Long Beach, California
	Fresno	MAR, JUNE, SEP, DEC	San Francisco, California
CONNECTICUT	Hartford	JAN, APR, JUL, OCT	Boston, Massachusetts
FLORIDA	Jacksonville	APR, OCT.	Miami, Florida
GEORGIA	Albany	FEB, AUG	Atlanta, Georgia

EXCITING WORLD OF HAM RADIOS 25

GUAM	Agana	JULY, SEP, NOV, JAN, MAR, MAY	Honolulu, Hawaii
HAWAII	Hilo	AUG, NOV, FEB, MAY	Honolulu, Hawaii
	Lihue	SEPT, DEC, MAR, JUN	" "
	Wailuku	AUG, NOV, FEB, MAY	" "
IDAHO	Boise	APR, OCT	Portland, Oregon
	Pocatello	NOV, JUN	" "
ILLINOIS	Rock Island	FEB, MAY, AUG, NOV	Chicago, Illinois
INDIANA	Fort Wayne	FEB, MAY, AUG, NOV	Chicago, Illinois
	Indianapolis	JAN, APR, JUL, OCT	" "
IOWA	Des Moines	MAR, JUN, SEP, DEC	Kansas City, Missouri
KANSAS	Wichita	MAR, SEP	Kansas City, Missouri
KENTUCKY	Louisville	MAR, JUN, SEPT, DEC	Chicago, Illinois
LOUISIANA	Shreveport	APR, OCT	New Orleans, Louisiana
MAINE	Bangor	FEB, AUG	Boston, Massachusetts
	Portland	MAY, NOV	" "
MICHIGAN	Grand Rapids	JAN, APR, JUL, OCT	Detroit, Michigan
	Marquette	MAY, NOV	St. Paul, Minnesota
MINNESOTA	Duluth	APR, OCT	St. Paul, Minnesota
MISSISSIPPI	Jackson	JUN, DEC	New Orleans, Louisiana
MISSOURI	St. Louis	FEB, MAY, AUG, NOV	Kansas City, Missouri
MONTANA	Billings	APR, OCT	Seattle, Washington
	Helena	APR, OCT	" "
NEBRASKA	Omaha	JAN, APR, JUL, OCT	Kansas City, Missouri

STATE	CITY	MONTH IN WHICH EXAMINATION ADMINISTERED	OFFICE ADMINISTERING EXAMINATION
NEVADA	Las Vegas	JAN, JUL	Long Beach, California
	Reno	APR, OCT	San Francisco, California
NEW MEXICO	Albuquerque	APR, OCT	Denver, Colorado
NEW YORK	Albany	MAR, JUN, SEP, DEC	New York, New York
	Syracuse	JAN, APR, JUL, OCT	Buffalo, New York
NORTH CAROLINA	Wilmington	MAY, NOV	Norfolk, Virginia
	Winston-Salem	1977-AUG, NOV	" "
		1979-FEB,APR,JUN,AUG,OCT,DEC	
	Charlotte	JAN, JULY	" "
NORTH DAKOTA	Bismarck	APR, OCT	Kansas City, Missouri
	Fargo	JUN, DEC	" "
OHIO	Cincinnati	FEB, MAY, AUG, NOV	Detroit, Michigan
	Cleveland	MAR, JUN, SEP, DEC	" "
	Columbus	JAN, APR, JUL, OCT	" "
OKLAHOMA	Oklahoma City	JAN, APR, JULY, OCT	Dallas, Texas
	Tulsa	FEB, MAY, AUG, NOV	" "
OREGON	Medford	SEP, MAY	Portland, Oregon
PENNSYLVANIA	Pittsburgh	FEB, MAY, AUG, NOV	Buffalo, New York
	Wilkes-Barre	MAR, SEP	" "
SOUTH CAROLINA	Columbia	MAY, NOV	Atlanta, Georgia
SOUTH DAKOTA	Rapid City	MAY, NOV	Denver, Colorado
	Sioux Falls	APR, OCT	Kansas City, Missouri
TENNESSE	Knoxville	MAR, JUN, SEP, DEC	Atlanta, Georgia
	Memphis	JAN, APR, JUL, OCT	" "
	Nashville	FEB, MAY, AUG, NOV	" "

EXCITING WORLD OF HAM RADIOS

TEXAS	Austin	**APR. OCT**	Houston, Texas
	Corpus Christi	**MAR. JUNE, SEPT. DEC**	" "
	El Paso	**JUN. DEC**	Dallas, Texas
	Lubbock	**MAR. SEP**	" "
	San Antonio	**FEB. MAY. AUG. NOV**	Houston, Texas
UTAH	Salt Lake City	**MAR. JUN. SEP. DEC**	San Francisco, California
VERMONT	Burlington	**MAR. SEP**	Boston, Massachusetts
VIRGINIA	Salem	**MAR. SEP**	Norfolk, Virginia
WASHINGTON	Spokane	**FEB. MAY. AUG. NOV**	Seattle, Washington
WEST VIRGINIA	Charleston	**MAR. JUN. SEP. DEC**	Detroit, Michigan
WISCONSIN	Milwaukee	**MAR. JUN. SEP. DEC**	Chicago, Illinois
WYOMING	Casper	**MAY. NOV**	Denver, Colorado

Figure 1-1b

ADDRESSES OF FCC FIELD OFFICES

Listed below are the addresses and telephone numbers of the FCC field offices. This list is alphabetical by state, and also includes the field offices in Puerto Rico and The District of Columbia (Washington, D.C.).

ALASKA, Anchorage U.S. Post Office Building, Room G63 4th & G Street, P.O. Box 644 Anchorage, Alaska 99510 Phone: Area Code 907-265-5201	FLORIDA, Miami 919 Federal Building 51 S.W. First Avenue Miami, Florida 33130 Phone: Area Code 305-350-5541
CALIFORNIA, Long Beach 3711 Long Beach Blvd. Suite 501 Long Beach, California 90807 Office Examinations (Recording) Phone: Area Code 213-426-7886 Other Information Phone: Area Code 213-426-4451	FLORIDA, Tampa 809 Barnett Bank Building 1000 Ashley Street Tampa, Florida 33602 Office Examinations (Recording) Phone: Area Code 813-228-2605 Other Information Phone: Area Code 813-228-2872
CALIFORNIA, San Diego Fox Theatre Building 1245 Seventh Avenue San Diego, California Office Examinations (Recording) Phone: Area Code 714-293-5460 Other Information Phone: Area Code 714-293-5478	GEORGIA, Atlanta Room 440, Massell Building 1365 Peachtree St. N.E. Atlanta, Georgia 30309 Office Examinations (Recording) Phone: Area Code 404-881-7381 Other Information Phone: Area Code 404-881-3084
CALIFORNIA, San Francisco 323A Customhouse 555 Battery Street San Francisco, California 94111 Office Examinations (Recording) Phone: Area Code 415-556-7700 Other Information Phone: Area Code 415-556-7701	GEORGIA, Savannah 238 Federal Office Building and Courthouse 125 Bull Street, P.O. Box 8004 Savannah, Georgia 31402 Phone: Area Code 912-232-4321 ext. 320 Other Information Phone: Area Code 415-556-7701
COLORADO, Denver Suite 2925, The Executive Tower 1405 Curtis Street Denver, Colorado 80202 Office Examinations (Recording) Phone: Area Code 303-837-4053 Other Information Phone: Area Code 303-837-5137	HAWAII, Honolulu 7304 Prince Jonah Kuhio Kalanianaole Building 300 Ala Moana Blvd. Honolulu, Hawaii 96813 Phone: Area Code 808-546-5640
DISTRICT OF COLUMBIA (Washington, D.C.) 1919 M Street N.W., Room 411 Washington, D.C. 20554 Phone: Area Code 202-632-8834	ILLINOIS, Chicago 3935 Federal Building 230 South Dearborn Street Chicago, Illinois 60604 Office Examinations (Recording) Phone: Area Code 312-353-0197 Other Information Phone: Area Code 312-353-0195
LOUISIANA, New Orleans 829 F. Edward Hebert Federal Building 600 South Street New Orleans, Louisiana 70130 Phone: Area Code 504-589-2094	NEW YORK, Buffalo 1307 Federal Building 111 W. Huron Street at Delaware Ave. Buffalo, New York 14202 Phone: Area Code 716-842-3216

MARYLAND, Baltimore George M. Fallon Federal Building Room 819, 31 Hopkins Plaza Baltimore, Maryland 21201 Office Examinations (Recording) Phone: Area Code 301-962-2727 Other Information Phone: Area Code 301-962-2728	NEW YORK, New York 201 Varick Street New York, New York 10014 Office Examinations (Recording) Phone: Area Code 212-620-3435 Other Information Phone: Area Code 212-620-3437
MASSACHUSETTS, Boston 1600 Customhouse 165 State Street Boston, Massachusetts 02109 Office Examinations (Recording) Phone: Area Code 617-223-6608 Other Information Phone: Area Code 617-223-6609	OREGON, Portland 1782 Federal Office Building 1220 S.W. 3rd Ave. Portland, Oregon 97204 Office Examinations (Recording) Phone: Area Code 503-221-3097 Other Information Phone: Area Code 503-221-3098
MICHIGAN, Detroit 1054 Federal Building & U.S. Courthouse 231 W. Lafayette Street Detroit, Michigan 48226 Office Examinations (Recording) Phone: Area Code 313-226-6077 Other Information Phone: Area Code 313-226-6078	PENNSYLVANIA, Philadelphia 11425 James A. Byrne Federal Courthouse 601 Market Street Philadelphia, Pennsylvania 19106 Office Examinations (Recording) Phone: Area Code 215-597-4410 Other Information Phone: Area Code 215-597-4411
MINNESOTA, St. Paul 691 Federal Building 316 N. Robert Street St. Paul, Minnesota 55101 Office Examinations (Recording) Phone: Area Code 612-725-7819 Other Information Phone: Area Code 612-725-7810	PUERTO RICO, Hato Rey (San Juan) Federal Building & Courthouse, Room 747 Avenida Carlos Chardon Hato Rey, Puerto Rico 00918 Phone: Area Code 809-753-4567 or Phone: Area Code 809-753-4008
MISSOURI, Kansas City 1703 Federal Building 601 East 12th Street Kansas City, Missouri 64106 Office Examinations (Recording) Phone: Area Code 816-374-5526 Other Information Phone: Area Code 816-374-6155	TEXAS, Beaumont Room 323 Federal Building 300 Willow Street Beaumont, Texas 77701 Phone: Area Code 713-838-0271 ext. 317
TEXAS, Dallas Earle Cabell Federal Bldg. Room 13E7, 1100 Commerce Street Dallas, Texas 75242 Office Examinations (Recording) Phone: Area Code 214-749-3243 Other Information Phone: Area Code 214-749-1719	VIRGINIA, Norfolk Military Circle 870 North Military Highway Norfolk, Virginia 23502 Office Examinations (Recording) Phone: Area Code 804-461-4000 Other Information Phone: Area Code 804-441-6472
TEXAS, Houston 5636 Federal Building 515 Rusk Avenue Houston, Texas 77002 Office Examinations (Recording) Phone: Area Code 713-226-4306 Other Information Phone: Area Code 713-226-5624	WASHINGTON, Seattle 3256 Federal Building 915 Second Ave. Seattle, Washington 98174 Office Examinations (Recording) Phone: Area Code 206-442-7610 Other Information Phone: Area Code 206-442-7653

Figure 1-1 (c)

With the Novice license you will be able to use telegraphy (A1) on the 40 and 80 meter amateur bands. The Technician's license will allow you to use telegraphy (A1) and voice on 80 m, 40 m, 15 m, and all higher frequencies. Figure 1-2 lists the frequencies of each band with the privileges allowed for each of the five license categories. The white areas are telegraphy (A1) only. The lightly shaded areas are telegraphy and voice. The heavily shaded areas are telegraphy, voice, and slow-scan TV. With your Novice license, your transmitter power may be 50 times what a CB operator can use—or 250 watts. With any of the more advanced licenses, your transmitter power may be 1000 or 2000 watts single sideband.

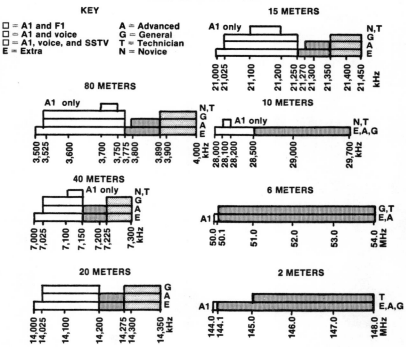

Figure 1-2 *Various amateur bands in meters and frequencies.*

As can be seen in Figure 1-2, the 80 meter band covers from 3500-4000 kHz; the 40 m band from 7000-7300 kHz; the 20 m band from 14,000-14,350 kHz; the 15 m from 21,000-21,450 kHz; the 10 m from 28,000-29,700 kHz; the 6 m from 50-54 MHz; and the 2 m from 144-148 MHz. The higher frequency bands starting from 2200 MHz and up are also listed. The 420-450 MHz band is used for amateur television (described later in this chapter) and the higher frequencies are useful for extremely short range experimentation. kHz and MHz denote frequency, while m denotes meters or wavelength. These distinctions are explained in detail at the end of this chapter.

SPECIALIZED COMMUNICATION

Amateur radio is not merely limited to telegraphy and voice transmission. High speed, conventional, and slow-scan TV transmission are also possible. In addition, phone patching, radioteletype, and satellite communication are exciting aspects of amateur radio. In Figure 1-2 you will see the frequencies for SSTV. Radioteletype is allowed on all frequencies. Satellite communication frequencies are determined by the equipment aboard the satellite.

AMATEUR TELEVISION (TV)

Long before commercial TV became popular, amateurs were sending and receiving crude television pictures over the air waves. They used motor-driven scanning and their efforts were doomed to failure. Those early mechanical systems were not too practical. In 1928 a QST article appeared on an all-electronic TV using the then rare-and-expensive cathode-ray tube.

By 1931, the cathode-ray tube was still a laboratory curiosity. In 1940, a few advanced amateurs were able to build TV transmitters using moderately priced iconoscopes. After World War II, the military surplus market gave amateur TV a new impetus. By 1960, amateurs were transmitting color TV signals. Ordinary home TV receivers are generally used, with their receiving frequency converted to the 420 MHz band.

One simple ham TV system uses an ordinary TV receiver, tuned to any TV station, with its video data blanked out but its raster information used. A transparency is placed on the face of a flying spot scanner. Light from the raster passes through the transparency, is picked up by a simple photoelectric tube, and fed to a video amplifier. This signal then modulates the radio transmitter. Sound and video are both sent on the same channel. These TV signals can then be received on any home TV receiver equipped with a 420 MHz converter. This system can be used to transmit movies. The movie film-projector light source is removed and a photo multiplier (such as a 931-A) installed in its place.

Another amateur TV system makes use of inexpensive closed circuit TV cameras which feed directly into the antenna terminals of a TV receiver. These TV cameras produce a high frequency video signal on TV channels 2, 3, 4, or 6. This video-modulated TV signal is fed into a mixer which changes the channel 2, 3, 4, or 6 signal into the 420 MHz amateur TV band. The 420 MHz signal is then amplified and fed to an antenna. From a good location, this simple TV station can have a range of 20 miles.

SLOW-SCAN TELEVISION (SSTV)

Amateur TV (ATV) is only for use on 420 MHz where its range is limited to less than 25 miles. Thus it is useful only to urban amateurs. For amateurs in sparsely populated areas, and those who would like to send TV pictures across the Atlantic ocean, there is yet another mode of amateur TV. This is SSTV. Its narrow bandwidth allows its use on any amateur band except 160 m. Essentially, video tones from black to grey to white are converted into specific tone or pitch. The vertical and horizontal synchronization pulses become another tone. Tones that represent picture information are interspersed with ordinary voice transmissions. As SSTV is merely a series of ordinary tones, it can be recorded on any tape recorder. The only limitation of SSTV is its slow frame rate. This means it takes 8 seconds to send one picture. On the other hand, any amateur band can be used. On 20 meters (14 MHz), SSTV signals are being sent around the world, as can be seen from the SSTV photos in Figure 1-3. Present experiments in sending

SSTV in color consist of sending three separate frames of the same picture. A red, blue, and green filter are successively placed in front of the camera lens for each of the frames. At the receiving end, corresponding filters are used and each frame is photographed on color film.

Figure 1-3 *Replica of first slow scan TV picture sent from Greenland by Amateur Radio station OX3LP. The polar bear is the national symbol.*

PHONE PATCHING

When you would like to talk to a friend in a distant city, you would use your amateur radio set and a phone patch. When you would like to make a phone call from your car, you would use your mobile amateur rig and a phone patch. A phone patch is a connection between your amateur equipment and the telephone line of your home phone. Phone patches have provided vital communication during natural disasters when telephone circuits have been disrupted. Phone patching allows amateurs to perform a great many public services for their communities.

Patching is accomplished by means of a coupler. A telephone line is usually a single pair of wires used for transmissions in both directions. A phone patch or coupler is connected to this single pair of wires. It connects the radio receiver's output circuit and the radio transmitter's audio input circuit to the telephone line. In other words, stations picked up by the receiver are sent through the telephone line, and a voice on the telephone line modulates the amateur transmitter.

An interesting phone patch application is the repeater phone patch. Chapter 8 will describe the popular amateur repeater. This is a receiver-transmitter located atop a mountain or skyscraper which picks up weak signals and retransmits them from an excellent location. With a repeater phone patch, a walkie-talkie with a touch-tone pad could direct dial any touch-tone telephone in the country. An amateur set in a car could do the same thing using a repeater phone patch.

SATELLITES

In Chapter 3, you will read about OSCAR (Orbiting Satellite Carrying Amateur Radio) 6 and 7. These two satellites orbit the earth at a speed of 16,000 miles per hour at a height of 900 miles above the ground. They have been placed there by a joint amateur-NASA program called AMSAT (Amateur Satellite Corp.), a nonprofit group supported by hams and the U.S. Government to encourage space communication experimentation. A satellite is nothing more than a repeater with orbital wings. It receives your tiny signal from earth and retransmits it over another amateur band. The curve in Figure 1-4 shows how the satellite's height above ground determines the possible range between a transmitter and receiver using the satellite. The time duration that any satellite will be within your range depends on that satellite's altitude. Higher altitude orbits provide longer exposure to the satellite. Also the longest time duration for a given altitude will occur on orbits which pass directly over your station's location. When you transmit to a satellite, you will be able to listen to your own signal being returned to earth.

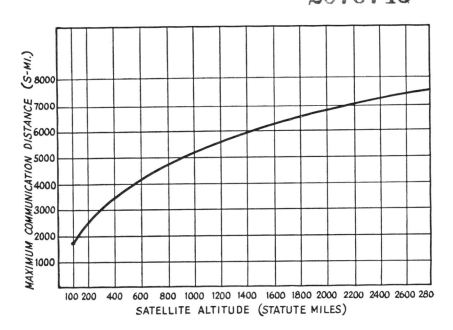

Figure 1-4 *Maximum communications distance vs. orbital height.*

Amateur satellites are usually smaller in size than their commercial counterparts. OSCAR, the Peace satellite launched in 1978, is no larger than 2 feet in height. Figure 1-5 shows OSCAR 8 with its complex array of solar cells and its various antenna.

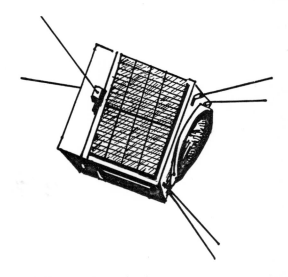

Figure 1-5 *OSCAR 8 successfully launched March 5, 1978. The various antenna and maze of solar cells can be seen.*

RADIOTELETYPE (RTTY)

After the typewriter became a common business instrument, it became obvious that there was a great need to send typewritten information over wires, and the teletypewriter was soon invented. This device allowed an individual in New York to type a written page, and at the same time, a teletype operator could receive the typewritten page in California. Teletype signals were transmitted over ordinary telephone wire; hence the name teletype. Daily newspapers needed vast quantities of typewritten data from all corners of the world, but overseas telephone lines were extremely expensive. It was not long before teletype signals were put on a radio transmitter in New York and received via short wave in Europe. The age of radioteletype was born.

Today thousands of radio amateurs have their own radioteletype stations sending printed pages of type to distant corners of the earth. The availability of inexpensive military surplus and used commercial teletype machines has undoubtedly given great impetus to amateur radioteletype. In Chapter 4 you will learn about the various ways to impress sound on a radio transmitter or to modulate a transmitter.

In radioteletype, a very unique technique of modulation is used. It is called Frequency Shift Keying, or simply FSK. Two distinct frequencies are transmitted which produce two separate tones. These tones correspond to a particular letter of the alphabet under the Murray encodement system. Figure 1-6 shows the Murray encodement for the letter F. Translating this into audiotones, we would have a high tone (1), a low tone (2), two high tones (3 and 4), and a low tone (5). The high tone is called mark and the low is called space.

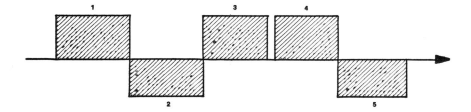

Figure 1-6 *Radio-Teletype uses the Murray encodement system to represent numerals and the alphabet. Using this system, this is how the letter "F" would appear.*

In the radioteletype receiver, a circuit called a demodulator (terminal unit-TU) changes the incoming mark and space RTTY signals into DC pulses. The DC pulses, in turn, activate (or key) selector magnets housed in the teleprinter which then move the corresponding typing hammers to make carbon impressions on a paper roll. The typing and printing mechanisms are connected through an electrical circuit called a loop.

It is easy to listen to teletype signals from distant continents. Tune a short wave receiver between 7 and 14 MHz where international short wave transmissions can be heard. Listen for brief tones rapidly switching back and forth. It will almost sound like a repetitive musical tone. This is the sound of RTTY.

FREQUENCY

A battery will give us direct current (DC)—that is, the flow of electrons is always in the same direction. There is always a positive battery terminal (+) and a negative battery terminal (-). Also the amplitude is constant—1-1/2 volts, 6 volts, 12 volts, etc. Alternating current (AC) consists of the sine wave in Figure 1-7.* Electrons move first in one direction, reach a positive peak voltage, then reverse, reach a negative peak voltage, then return to zero volts. This is one AC cycle. A cycle travels in time. In Figure 1-8 two cycles occur in a second. Thus the rate or frequency is two cycles per second, or 2 hertz. The frequency of sound is 20-15,000 hertz, and 15,000 can be expressed as 15 kilohertz. Radio waves have a frequency from 0.2 to 900 million hertz. This can be expressed as 0.2 to 900 megahertz. For example, the CB band is a range of frequencies from 27 to 27.4 million cycles per second, or megahertz.

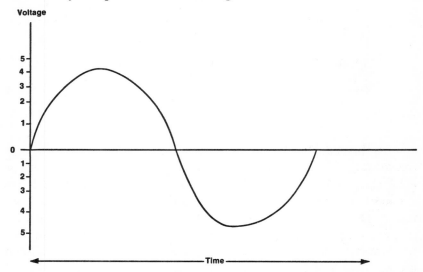

Figure 1-7 *The AC sine wave. The x axis (horizontal) represents time, the y axis (vertical) represents amplitude or voltage. Everything above the zero axis is positive, and below is negative.*

---

*A sine wave reaches a positive peak, then reverses itself, reaching a negative peak. It is called a sine wave because its instantaneous position in degrees is defined by the sine of an angle formed by the wave's base with a line drawn to the instantaneous position. A vertical line from the instantaneous position to the base (hypotenuse) forms the opposite side of a right triangle. A completed sine wave is 360°.

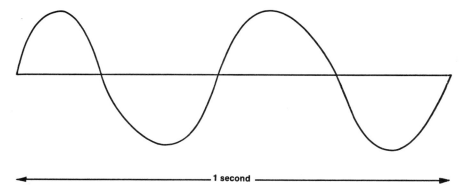

Figure 1-8 *An AC wave. Two complete cycles occur in one second. Thus we can say the rate or frequency of this wave is 2 cycles per second or 2 hertz.*

WAVELENGTH

We have all heard a super-jet breaking the sound barrier or traveling faster than the speed of sound. However, we will never see a super-jet breaking the light barrier by traveling faster than 186,000 miles per second. Light waves and radio waves both travel at about the same speed; and 186,000 miles a second is the equivalent of 300,000,000 meters a second.

If a radio wave has a frequency of 300 megahertz (million cycles per second), one cycle will move one meter (39 inches) before the next cycle begins, and a three megahertz wave moves 100 meters before its next cycle begins.* As the frequency (in megahertz) decreases, the wave's 'space' (in meters) increases. This relationship is expressed by the formula:

$$\text{wavelength in meters} = \frac{300}{\text{frequency in megahertz}}$$

$$\frac{300}{15 \text{ megahertz}} = 20 \text{ meters}$$

---

*These two waves do not travel at different speeds. Their speed is identical, but the length of each wave is different.

METERS AND MEGACYCLES OVER THE RADIO FREQUENCY SPECTRUM

In the earliest days of radio, most attempts at communication were made in the area of 200 meters. This portion of the radio frequency spectrum is now known as the AM radio broadcast band. Several years later, the crude spark-gap transmitters were operating on a relatively high frequency, around 100 meters. It was discovered that long distance communication was far easier to achieve at 100 meters than it was at 200 meters. This was not surprising for 100 meters is only slightly lower in frequency than the popular amateur 80 meter band. Over the frequency range of 3.5 to 4 MHz, distant communications can be achieved even with the use of low power transmitters.

Of course, as the frequencies move up, their designation in meters moves down. At frequencies above many of the popular amateur bands we find the FM broadcast band.

This is 88-108 MHz, or approximately 4 meters. The commercial TV spectrum starts at about 5 meters (60 MHz) and goes all the way (in the UHF region) up to about 1 meter. In the UHF (ultra-high frequency region), there is an amateur band called 70 centimeters (430 MHz). Its wavelength in space is of course shorter than a meter.

Moving even higher in frequency into the shorter centimeter wavelengths, we approach the microwave region. The invention of microwave tubes such as the klystron and the magnetron led to the perfection of radar. Luckily, a crash program in 1942 made radar a reality. Consequently, Great Britain could repel and defeat the Luftwaffe (German Air Force). We are all familiar with microwave cooking. Small magnetrons (microwave oscillators) generate microwave energy that heats the water molecules in food.

As we advance beyond the microwave region, we find the infrared wavelengths, then the visible light spectrum, and up into the ultraviolet region. At these wavelengths, we no longer use the terms meter or centimeter to describe the wave. Its extreme short wavelength necessitates the use of the term *angstrom*. An angstrom is in the order of 10^{-6}, or a millionth of a centimeter. Moving into even shorter wavelengths than angstroms, we find *Alpha, Gamma,* and *X-Rays.* In this range, the characteristic wave front looks so different that it no longer resembles a wave. Atoms begin to break down into their parts or particles. This is why energy in this part of the spectrum is called particle radiation.

2

Understanding the Ham Bands and Propagation

RADIATION FROM THE SUN

In the transmission of radio waves over short distances, the signal usually travels from transmitter to receiver in a straight line. However, when radio waves cover long distances, their transmission path or propagation can be quite complex. They may bounce back to earth from charged particles in the atmosphere one or more times. Charged particles congregate together in layers known as regions of ionization. As we know, what happens in the upper atmosphere is largely determined by solar activity.

One good index of solar activity is sunspots, or the solar flux. The National Bureau of Standards in Boulder, Colorado uses its radio station WWV to broadcast solar flux from the measurements taken several times a day. These broadcasts can be heard on any short wave receiver 19 minutes after each hour on 5, 10, and 15 megahertz. The solar flux index is derived from a microwave receiver measurement and is directly related to a count of sunspots known as the Zurich sunspot number. The Zurich number technique, with some modifications, has been in use for over 200 years. Figure 2-1 shows the relationship between Zurich sunspot numbers and the solar flux index broadcast by WWV.

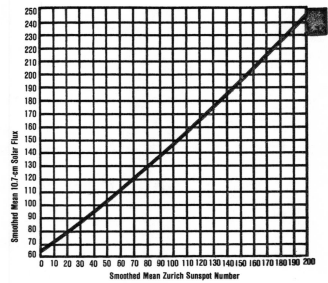

Figure 2-1 *Zurich sunspot number vs. solar flux received on a microwave receiver.* **Courtesy ARRL**

SUNSPOTS

Since solar activity is a key factor in long distance radio communication, we should be familiar with its leading index—sunspots. A sunspot is an actual hole in the solar surface brought about by a swirling magnetic vortex which transports solar atmospheric hydrogen back into the sun's core. The origin of sunspots is still something of a mystery. However, we know that they have several cyclical patterns.

Short term sunspot cycles are determined by the rotational period of the sun. This varies from 22 to 28 days because the sun is a nonsolid gaseous body. Its poles rotate much faster than the sluggish gasses near the solar equator. Thus sunspots near the equator will return to the same position on the solar surface (as seen from earth) every 28 days, while sunspots near the poles will return to the same position every 22 days. Sunspots affect radio propagation most when they are first observed on the solar surface. Consequently, short term changes in radio propagation take place every three to four weeks.

Many high school science programs currently use simple 10 to 30x power telescopes to observe sunspots. A solar filter is placed on the telescope and its image is projected onto a black cardboard surface. In this manner, you could take your own sunspot count and develop your own solar flux index showing changing radio propagation conditions. It should be noted that solar radiation caused by sunspot activity can be transmitted as ultraviolet light and also as charged particles. Ultraviolet light will reach us eight minutes after the event, or the same time that we see it. Particles move more slowly and can take up to 40 hours to reach the earth. Particles can absorb radio energy, causing radio signals to weaken, but ultraviolet energy stimulates ionization, making radio signals stronger.

Sunspots also have long term cycles. These are related to the orbits of Jupiter, Mercury, and other planets. Peaks in these cycles occur every eleven years, every twenty-two years, and every 180 years. The next major peak will occur in 1981. During the last peak in 1959, European TV stations could be seen on TV screens all over the U.S. Taxicab radios in the U.S. could be heard by taxicabs in the Middle East. This was the result of a large number of sunspots which stimulated ionization in the upper ionosphere. As we shall discover, the ionosphere has several layers of ionization.

THE IONOSPHERE

As we can see in Figure 1-2, there are six different amateur bands below 30 megahertz or 10 meters. These relatively low frequency bands can only propagate radio signals over any distance when the correct conditions in the ionosphere exist. This is a region between 60 and 200 miles above the earth's surface. Free ions and electrons are found in great numbers in this region. Ultraviolet radiation creates several layers in the upper atmosphere. Each layer has a central region of dense ionization where the electrical charges are strong. Weakly ionized regions possess a weak electrical charge.

THE LAYERS OF IONIZATION

Figure 2-2 illustrates the various layers of ionization. The lowest layer of major importance is the E layer. This layer is only about 60 miles high and is a very influential region. The earth's atmosphere is still fairly dense at this height. Thus electrons and ions set free by solar radiation rapidly recombine to form particles which are electrically neutral. The result of this particle recombination is that the E layer will bend radio waves back to earth, as can be seen in Figure 2-2. Consequently, the E layer will only bend radio waves while the sun is present. Ionization, or bending by the E layer, is greatest near noon and nonexistent after sundown or during the night.

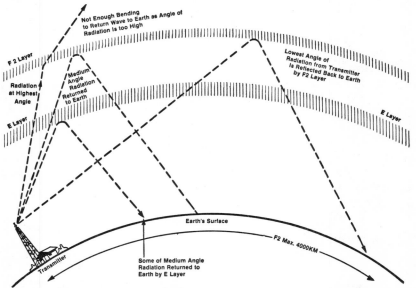

Figure 2-2 *How different ionosphere layers affect the reflection of radio waves.*

Another less important region of ionization is the D region, which is only 40 miles high. The density of this layer is proportional to the height of the sun. The D layer absorbs signals on 1.8 MHz (160 meters) and 3.5 MHz (80 meters) so that propagation only takes place on these bands during daylight hours for a very short distance.

As we go higher up into the atmosphere, to approximately 140 miles, we come to the F layers. Here the air is so thin that recombination of ions and electrons occurs very slowly. It begins to decrease after sundown and is at its minimum just before sunrise. During the daytime, the F layer has two parts, F1 and F2. F1 is 140 miles high, while F2 is 200 miles high. Both F1 and F2 come together at sunset.

As a general rule, the more intense the ionization of a particular layer, the more it bends the radio wave. The frequency or wavelength of the signal reaching the layer also affects the amount of bending that takes place. If we assume that an ionized layer has a fixed density, then longer wavelengths (lower frequency signals) are usually bent more. It then follows that the lower frequency amateur bands will allow us to have longer periods of long distance communication. Remember that the intensity of solar radiation is always a factor when considering the ability of the ionized regions to deflect radio waves. Increased solar radiation produces increased ionization, hence increased bending.

Another factor that must be considered when discussing reflection from the ionized regions is absorption. When radio signals reach the ionized regions, there is a collision of particles. In this collision, friction and heating result in a loss of energy. More important, the greater the density of the atmosphere, the more oxygen is present and the more heating occurs. Thus waves that are bent by the E layer (60 miles up) will lose more energy than waves bent by the F layers (140-200 miles up), where the atmosphere is thin.

The various factors involved in determining the amount of loss occurring when radio waves are bent back to earth need not be considered when one simple rule is applied. To minimize loss, always use a frequency close to the maximum frequency that will be bent back to earth at a given time. For example, if all frequencies above 8 MHz are not deflected by an ionized layer, then the frequencies between 7 and 8 MHz will be deflected with the least amount of loss by absorption. This highest possible frequency of deflection is called the *critical frequency*. Different layers of ionization can have different critical frequencies. This explains why relatively higher frequency amateur bands (20 meters, 15 meters, 10 meters) can propagate very strong signals over extremely long distance paths, seldom possible on the lower frequencies.

To illustrate this, Figure 2-3 shows a 21 MHz (15 meter) signal which is above the critical frequency of the E layer (dense atmosphere) and is thus not deflected by that layer. However, at that particular time, 24 MHz is the critical frequency of the F1 layer. Therefore, being below this critical frequency, the 21

MHz radio wave is deflected back to earth by the F1 layer. If it were bent by the lower E region at the same deflection angle, the wave's propagation distance would only be half what it is. Also, deflection by a layer where the atmosphere is less dense produces less loss, returning a stronger signal to earth.

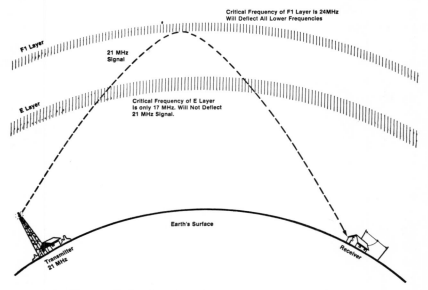

Figure 2-3 *Each layer has its own critical frequency.*

Another rule in propagation is that a given density of ionized region will bend more of the radio wave energy when the wave reaches this layer at the sharpest possible angle. In Figure 2-4, two different transmitters are used. Transmitter 1 radiates its signal at a high radiation angle. This signal does not reach the ionized region at a sharp angle and much of it is not deflected as it goes through the ionized region. However, transmitter 2 radiates its signal at a low radiation angle, reaching the ionized layer at a sharp angle. In this case, much more of the wave's energy is deflected by the ionized layer. Consequently, a far greater amount of energy is returned to earth.

The distance between the transmitters in Figure 2-4 and the point where the signals return to earth is called the *skip distance*. Skip also refers to the propagation of a wave whenever

HAM BANDS AND PROPAGATION

deflection from any ionized layer is involved. When there is no bending of a wave and it merely travels along the ground, it is known as a *ground wave*. Both skip and ground wave signals may occur at the same time. For example, if a CB operator in New York talks to a friend 10 miles away, he or she is talking over a ground wave path. At the same time, another CB operator in Texas may be listening to the New York CB operator over a 1000 mile skip path.

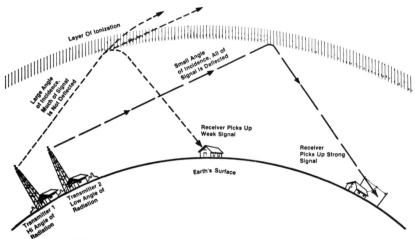

Figure 2-4 *Wave bending and the angle of incidence.*

PROPAGATION BULLETINS

WWV and WWVH are two stations operated by the National Bureau of Standards which gives hourly propagation data. Every hour, at nineteen minutes after the hour, both stations give the state of the earth's geomagnetic field, solar flux, and general propagation conditions. WWV can be heard on any shortwave radio at 5 MHz, 10 MHz, or 15 MHz. For additional material on the use of these bulletins, refer to QST articles in June, August, and September 1975.

THE LOW FREQUENCIES

The lowest frequency amateur band is 160 meters. Daytime distance on this band is 25 miles. On cold winter nights, European stations can be heard on this band. The 3.5 MHz band (80 meters) can be used for a 200 mile distance during the day, and a several thousand mile distance late at night.

The 7 MHz band (40 meters) is used for 500 mile contacts during the day. At night, signals will propagate on this band around the world.

The 14 MHz band (20 meters) can be used for 1000 to 3000 mile distances during daylight. In addition, during years when peak solar activity occurs, this band can also be used for long distance contacts at night.

The 21 and 28 MHz bands (15 and 10 meters) are most useful for long distance propagation during periods of intense solar activity and during peaks in the sunspot cycle. The remainder of the time they are useful primarily for ground wave.

THE HIGH FREQUENCIES

The VHF (very high frequency) and UHF (ultra-high frequency) part of the spectrum starts with the 6 meter band (50 MHz) and continues up through the 2 meter band and beyond, into the experimental amateur bands. Up until the 1960s, these bands were useful for ground wave distances up to 25 miles if considerable power was used and both stations had good locations. In recent years, these distances have been extended to 100 miles using very low power and only a fair location. This has been made possible by remote, unattended relay stations located on the top of a mountain or atop a skyscraper. These repeater stations will be discussed in detail in Chapter 8.

The next chapter covers a relatively new phenomenon for amateurs who operate the VHF and UHF bands. The advent of NASA launched, amateur satellites (OSCAR) means that repeater or relay stations are now orbiting the earth. OSCAR 6 and 7 receive and retransmit your signal, giving the high frequency operator the ability to talk around the world. Amateur satellites to be launched in 1979 and thereafter will enable solid contacts to be made around the world with the same consistency as talking to a station 10 miles away.

In the past, amateurs depended on the ionosphere to deflect radio waves back to earth. Although this is still the major technique of long distance communication, it depends on both the short and long term cycles of sunspot activity. In striving to establish higher reliability than ever before in international communication, the amateur satellite is truly creating the long heralded global village.

3

Talking to Europe on Your Walkie-Talkie

SPACE COMMUNICATION FOR EVERYONE

Many old time hams are returning to amateur radio because they have been bitten by the satellite bug. Some say the OSCAR amateur satellites give them the feeling of being in the 21st century.

When the Russians launched Sputnik I, the age of the satellite was suddenly upon us. In 1962, a small group of amateurs designed and built OSCAR I, the world's first nongovernmental satellite. Like the six satellites that have followed, it was built of, by, and for the radio amateur. There has since been unprecedented enthusiasm over the satellites. Today's ham satellites travel 15,000 miles per hour, 910 miles over our heads. In 1979, the Amateur Satellite Corporation (AMSAT) will launch OSCAR (Orbiting Satellite Carrying Amateur Radio) Phase III (Figure 3-1). This revolutionary device will allow hams in the United States to talk to Europe and Japan at the same time for six hours every day. Its disaster relief and medical and educational benefits alone will be enormous.

THE HISTORY OF AMATEUR SATELLITES

The first five satellites (Phase I) laid the groundwork for the more sophisticated satellites. Amateurs learned to track an orbiting satellite, receive and interpret telemetry, and use this data to command the spacecraft. Phase II began with the launch of AMSAT-OSCAR 6, the first long-lifetime, solar-powered amateur satellite. Transponders aboard these satellites automatically receive earth stations on one frequency band, then retransmit them back to earth on another frequency band. Experiments with Phase II satellites have shown that they can perform feats besides amateur radio. They can pinpoint downed aircraft from emergency locator transmitters. They can relay an electrocardiogram from a moving auto to a distant hospital. The block diagram in Figure 3-2 shows equipment aboard a typical satellite.

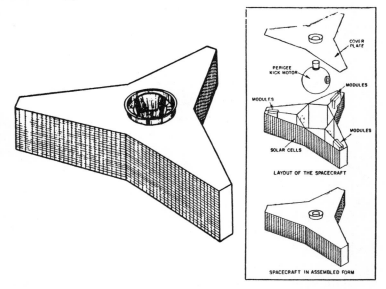

Figure 3-1 *Thanks to its elliptical orbit which will take it nearly 40,000 km into space, the Phase III OSCAR will enable amateurs to relay medical data or other information over thousands of miles for up to 15 hours at a time. Scheduled launch is late in 1979.*
Phase III Satellite: The start of worldwide satellite DX.

EQUIPMENT REQUIRED TO TALK TO OSCAR

As of 1977, the most intriguing satellite in orbit is AMSAT-OSCAR 7. Of its two modes of operation, Mode B is the most interesting. In this mode, as we see in Figure 3-3, it receives on

EUROPE ON YOUR WALKIE-TALKIE

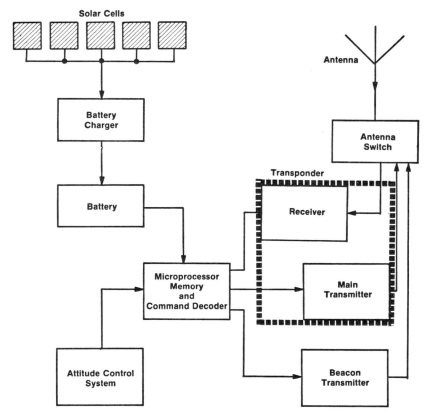

Figure 3-2 *Simplified version of equipment aboard a typical satellite.*

432 MHz, or 70 centimeters, and sends on 145 MHz, or 2 meters. The phenomenal sensitivity of the 432 MHz receiver has surprised even its designers and makes possible regular satellite communications with only a few watts of power. OSCAR 7, Mode B, is by far the preferred function by experienced satellite communicators. It far surpasses OSCAR 6 in sensitivity and downlink signal strength, and is in turn superior to OSCAR 7, Mode A in the same areas.

Almost any receiver capable of tuning the proper segment of the two-meter band, or 145.9-146 MHz, will receive the downlink signals from OSCAR 7, Mode B. The signals are much stronger than those from OSCAR 6, which can be heard on 10 meters, between 29.45-29.55 MHz. The satellite beacon transmits an easily recognizable, continuous dit-dah. The OSCAR 6 beacon is on 29.45 MHz, while OSCAR 7 beacon, Mode A, is 29.40 MHz and Mode B is 145.975 MHz (see Figure 3-3).

To receive satellites, the 10 meter or 2 meter receiver should have a beat-frequency oscillator (BFO). BFOs will be discussed in greater detail in the next chapter. Any sensitive receiver that can receive Morse code or single sideband (SSB) can be used. The 2 meter AM rigs can sometimes be purchased for $50 at auctions and flea markets. Many books describe how to build a simple BFO.

To transmit to OSCAR 7, Mode B, on 432 MHz (see Figure 3-3), transmitters which were used on the commercial 450-470 MHz band can be tuned down to 432 MHz. There are also amateur transmitters made specifically for 432 MHz. To send Morse code to a satellite, the transmitter need only be turned on and off with a telegraph key. Low price, low power transmitter kits for this band are available from VHF Engineering, 320 Water Street, Binghamton, NY, or from Hamtronics, 132 Belmont Road, Rochester, N.Y. A technique called "pulling" the crystal can be used to make these transmitters cover the entire satellite uplink band. Several manufacturers make single sideband transceivers. The KLM Echo transceiver has sufficient output for OSCAR 7, Mode B, operation.

A popular approach to satellite communications is the use of a transverter. These devices convert one frequency band to another. Texas RF Communications, 4800 West 34th Street, Suite D12A, Houston, Texas 77092, imports Microwave Module MMT432 transverters from England, which receive 38 MHz from a CW or SSB transmitter and convert it up to 432 MHz with 10 watts output. ARCOS, Box 546, East Greenbush, NY also markets a similar transverter. It is easy to tell if a satellite is being "triggered" by your transmitter. Simply listen on the downlink frequency to hear the slightly delayed signal you have just transmitted.

The OSCAR 7, Mode B, downlink on two meters can be heard with almost any two-meter antenna and an adequate receiver. The more elaborate the antenna, the more accurately the antenna must be pointed as it is sharply focused. Simple,

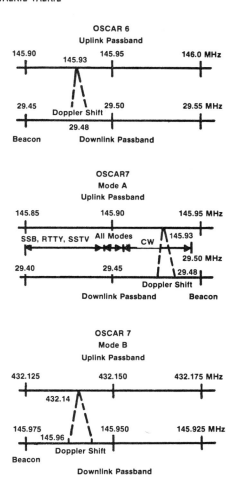

Figure 3-3 *Frequencies of OSCAR 6 & 7.*

low cost receiving antennas will thus make tracking much easier.

Making OSCAR 7, Mode B, ham contacts is similar to ordinary, nonsatellite amateur contacts, except that signals are much stronger. Often European stations outnumber U.S. stations during the satellite's mid-Atlantic passing. The same 432 MHz gear you use to talk to satellites can be used to talk to local amateurs directly. Care should be taken to avoid nearby frequencies where experiments are being conducted which bounce radio signals off the moon.

In trying to hear satellites, you will notice that their signals are weaker when they are near the horizon than during overhead passes. This is so because the distance to the satellite is greater. Path loss increases, and the greater distance through the lower atmosphere adds additional absorption. Finally, trees and other vegetation which have no effect on overhead passes introduce loss or attenuation in signal strength on passes close to the horizon. Considerate amateurs trigger satellites using the minimum possible transmitting power. Sending high power transmissions to a satellite forces it to retransmit higher power. This not only shortens the life of its battery, but also the life of the satellite itself.

DX or distant stations can be contacted with satellites, depending on the height of the satellite. For OSCAR 6 and OSCAR 7 at an altitude of 910 miles (1500 km), maximum range—with the satellite directly between you and your contact—is about 4,900 miles (7,800 km). A circle drawn on a map will show you what countries fall within this range. From the eastern United States, parts of six continents and well over 100 countries fall within range of possible contacts. Of course, not all these countries have active satellite communicators.

EUROPE ON A WALKIE-TALKIE

The low power needed to trigger OSCAR 7, Mode B, means that a satellite ground station can be held in the palm of your hand. A highly directional beam antenna can be pointed to the satellite. Such an antenna can be homemade and is described in Chapter 12.

SPACE EXPERIMENTS

There are two features of satellite operations of special interest. Every Wednesday, OSCAR 7 is reserved for special space experiments. These are coordinated by AMSAT Vice-President of Operations Rich Zwirko, K1HTV. Past experiments have included sending electrocardiograms from a moving vehicle to a hospital thousands of miles away. Canadians have used OSCAR for emergency transmitter locating, as in the case of a downed aircraft. Teletype can be sent over OSCAR. If you are interested in a computer linkup via amateur satellite, or have some other special experiment in mind for OSCAR 7, contact KIHTV, 36 Sweet Birch Drive, Meridien, Connecticut 06450.

Another aspect of satellite operation has created great excitement and holds fantastic potential for the future. This is the twice yearly linking experiments. Since OSCAR 6 and OSCAR 7 are in slightly different orbits, OSCAR 7 overtakes and passes its predecessor about twice each year. When these satellites are within several hundred miles of each other, a double satellite relay can be achieved. Figure 3-4 shows how this link-up is accomplished. The 432 MHz uplink to OSCAR 7 is translated to 145.9 MHz. This is picked up by OSCAR 6 and sent to ground stations at 29.45 MHz. The first OSCAR links in 1973 comprised the first free access satellite-to-satellite relays in the history of space flight.

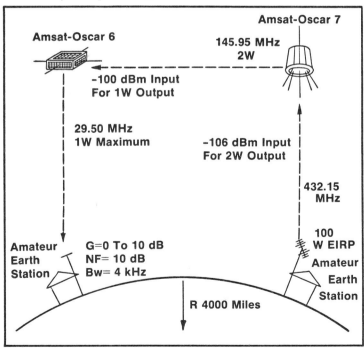

Figure 3-4 *Satellite-to-satellite links are possible out to a separation of 5000 km.*

An amateur license is not needed to monitor satellites during space experiments or at any other time. Information on signal strength, fading telemetry data, etc., can be sent to AMSAT to help them evaluate the success of these projects. Many high schools and colleges are now doing this. This mass of data is extremely useful to the amateur designers of future satellites.

Volunteer amateur command stations can actually control the various mechanical functions aboard a satellite. Randall Smith, VE3SAT is the volunteer command station operator for eastern North America. His completely automated station, operated by a microprocessor, provides continuous control over the satellite while it is over this part of the globe. AMSAT coordinators arrange for demonstrations and spread the word about operating problems or schedule changes. These coordinators are always looking for assistance. Write to ARRL, Newington, Connecticut 06111 for their names.

LOCATING OSCAR

You're listening to the hissing noise of 10 meters or 2 meters. Suddenly signals explode all over the band. South American, African, and European amateurs are on Morse code and SSB. The steady dit-dahs of the satellite's beacon are heard. OSCAR—Orbiting Satellite Carrying Amateur Radio—is headed your way.

How do you pinpoint a tiny metal box speeding 910 miles above the earth at 16,000 mph? The American Radio Relay League, Newington, Connecticut will supply any listener or amateur with a device called an OSCARLOCATOR. With it, you can predict when both active satellites (OSCAR 6 and 7) will be within range of any location in the Northern Hemisphere.

The OSCARLOCATOR has three parts—a polar projection of the earth and two circles on an acetate sheet. The larger circle with numbers around its circumference is the Orbit Finder, and the smaller is the QTH/Rangefinder. When the Rangefinder is centered on your location (QTH), it shows the satellite's signal range. The larger Orbit Finder is positioned

EUROPE ON YOUR WALKIE-TALKIE 61

over the map and turned so that the number 0 is on the longitude where the satellite crosses the equator on the day you are listening. Figure 3-5 shows both acetate sheets held in place over a globe whose center is the North Pole. The heavy line calibrated in minutes shows the satellite's Orbital Track on a

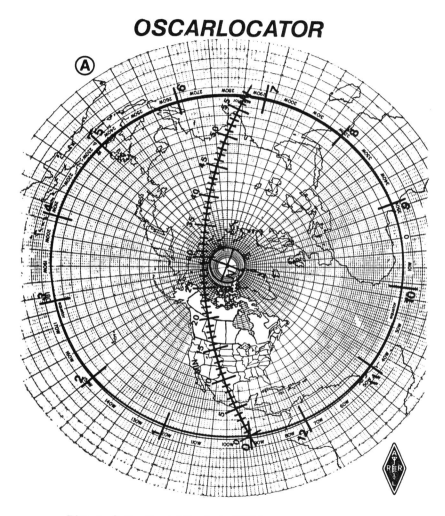

Figure 3-5 *Special thanks to K2ZRO.*
OSCAR's orbit for a given day. The precise time over a given location can be calculated.
Courtesy ARRL

particular day. In this case, your location is Houston and it is November 8, so that OSCAR 6 is crossing the Equator at 88° west longitude. A Time Conversion Chart (also supplied) changes Universal Coordinated Time (time at the zero or reference meridian) into your local time. In that way, you will know exactly when each satellite can be heard, and for how long, on a particular day.

THE PHASE III SERIES

The Phase II series of satellites that are in orbit today are available less than two hours a day for most amateurs. The creators of Phase III satellites, the first to be launched in 1979, will overcome this limitation. They considered the various solutions to the problem of limited satellite accessibility and came up with a solution.

A known fact is that more than 90 percent of the earth's amateur population is located in the Northern Hemisphere, and the amateur population is distributed, unevenly, over all geographic longitudes. All amateurs wish to talk to satellites for a maximum period of time each day and want to communicate over the maximum geographic separation, with no penalty for talking to relatively nearby stations. Based on these objectives, it was found that the Phase III satellites should meet the following criteria: they should spend a majority of their time in the Northern Hemisphere; they should be as high above the earth as is practical (allowing for the fact that signal strength decreases as the inverse square of the height); and they should not favor any geographic longitude. Popular orbits of the past, called geosynchronous, favor one geographic longitude, and are thus not desirable when considering all amateurs. With the help of Newton's laws of gravity and other scientific tools, a family of orbits was finally decided upon. Figure 3-6 shows three possible orbits for the Phase III series of satellites.

It would be impossible to locate a satellite permanently over the North Pole, and since 10 percent of the amateur population lives in the Southern Hemisphere, it cannot be neglected. Thus the orbits shown in Figure 3-6 are nearly perfect. This fact is proven by the formula which depicts the communications efficiency of an orbit.

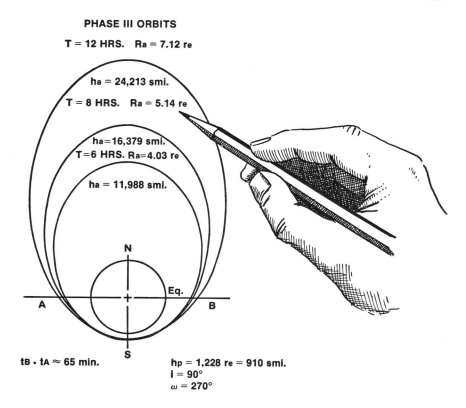

Figure 3-6 *Three possible orbits for the Phase III spacecraft are compared on this diagram. Notice that the time it would spend in the Southern Hemisphere remains nearly constant for each of the three orbits. The most elliptical orbit would thrust the spacecraft more than 24,000 miles into space at apogee, the farthest distance from earth.*

Communications efficiency =

$$\frac{\text{time space craft is above Northern Hemisphere}}{\text{total period of orbit}}$$

As the satellite is at its northernmost point, virtually every station in the Northern Hemisphere could communicate with every other—something that has never before happened in amateur radio. Using communications transponders in the 2 m and 70 cm bands, users of the satellite would never experience skip propagation or "no propagation" problems common to ionospheric communications. Even severe ionospheric disturbances would affect the performance of the satellite communicator only slightly.

THE TECHNICAL PROBLEM

Phase III spacecraft must be capable of injecting itself into the final desired orbit, and it must provide reliable communications for ground stations with simple equipment.

Unlike past OSCARs, the final desired orbit for Phase III cannot be obtained directly by having it launched piggyback on another space mission. The only method of obtaining these orbits is for the Phase III spacecraft itself to have a propulsion capability so that available orbits may be changed into an orbit of the type discussed.

Orbit modifications require that the satellite's velocity be increased at some point along its path. The velocity increase, or ΔV, may be thought of as a vector with a component along the orbit "track" and components perpendicular to the track or "crosstrack." Velocity added along the track will change the apogee and/or the perigee of the orbit while one of the crosstrack components will change the inclination of the orbit. Once the properties of the initial orbit and the final desired orbit are known, the necessary ΔV and firing angles may be calculated.

The two types of orbits available to AMSAT for Phase III launches are low altitude polar orbits and synchronous transfer orbits, with a high apogee, low perigee, and low inclination. The ΔV requirements to change both of these orbits to the Phase III orbit have been studied in detail. The velocity increase in each case is about the same (1,500 m/s). In one case, however, the "kick" of the propulsion system is applied along the track to increase the apogee, and in the other it is applied crosstrack.

AMATEURS AND PROFESSIONALS

As the satellite program has unfolded, amateurs and professionals within government and industry have increasingly worked together. Some professionals did not take amateurs too seriously in the past, but today these same professionals have much more respect for amateur volunteers in the satellite program. This has come about because certain technical problems which professionals have had difficulty with were solved by the amateurs who worked alongside them. Recently, a government official who represented a $50 million NASA spacecraft commented, "I have to know more about this ham hobby satellite before I allow it to piggyback on the next NASA satellite."

The government official was responsible for millions of dollars of the taxpayers' money. He deserved a serious review of the ham hardware flying with the NASA spacecraft. AMSAT has an excellent record in convincing the professional/technical community that amateurs are highly competent. As the Phase III launch gets underway, professional/amateur interaction becomes even more important. The kick motor that will be part of Phase III could be a hazardous system. In this challenging venture, a new breed of professional/amateur will be born.

THE MECHANICS OF PHASE III

Radio amateurs can now transmit to and receive high-quality signals from an orbiting spacecraft for up to 23 minutes at a time. How do we go about providing that same level of signal strength over a 12 hour elliptical orbit that will bring the satellite about ten times farther from its users? How can we determine the most useful and practical frequencies for the transponders and beacons aboard this revolutionary satellite?

In order to provide consistently strong signal levels at 20,000 miles as well as at 2,000, the effective radiated power (erp—apparent power level out of the antenna) of the Phase III spacecraft must be increased over that of AMSAT-OSCAR 7 to a level of 1,000 watts. As with ground stations, this can be achieved through a combination of increased transmitter power output and antenna gain.

The transmitter power output impacts dramatically the power system which must deliver the DC input. It also affects the thermal design of the satellite, since the additional heat produced must be conducted and radiated away from the final stages. In turn, the antenna affects the attitude control system; the higher the antenna gain, the more accurately it must be pointed toward the earth.

The antenna will be located on the spinning axis of the spacecraft and will be pointed directly toward the earth at apogee. The altitude of the spinning body will remain fixed throughout the orbit. As the satellite comes away from apogee, the distance to the user will decrease and the path loss will drop. The earth, however, is no longer centered in the beam, thus serving to decrease antenna gain. The antenna can be designed so that the reduced path loss is almost exactly counteracted by the decreased gain, resulting in a nearly constant signal level to the user during the entire pass.

To provide high-quality signals for the perigee part of the orbit, the transmitter will be switched to an omnidirectional antenna just north of the equator. It will switch back to the high-gain antenna on the outbound leg of the ellipse.

The Phase III user will gain access to the spacecraft with the same simple low-power equipment required for the earlier satellites. This remains an important objective of all future satellite programs.

BEST FREQUENCIES

As shown in Figure 3-7, the best bands for the Phase III communications transponders are 2 m and 70 cm. Ten meters was eliminated because antenna gain at the satellite would still be required, and such antennas would be very difficult to implement. The transponder bandwidth is limited to 155 kHz by an IARU Region I agreement regarding the satellite subband at 2 meters. This frequency band, from 145.845 MHz to 146,000 MHz, will be completely used by the communications system.

A beacon is a beep-type transmitter that is used for identification. Two beacons will be associated with each transponder, one at each end of the downlink passband. One will be a general beacon, providing bulletin information and low-speed telemetry of interest to users. The other will be a 400 bit-per-second engineering beacon which may be utilized by individuals with home microprocessor systems.

A 2 watt S-band beacon at 2304.1 MHz will be included to provide both Doppler measurements for orbit determination and a source of backup telemetry. This beacon will be operated on an as-needed basis near orbit apogee.

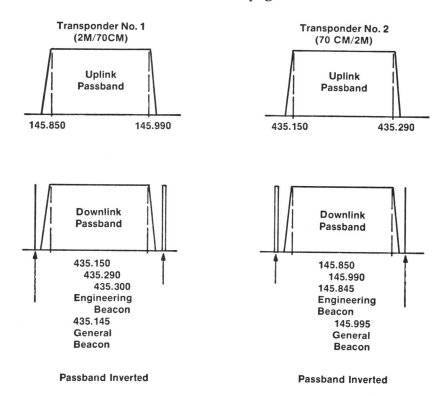

Figure 3-7 *Transponder frequencies of the Phase III spacecraft. Notice that each transponder will have two beacons.*

THE POWER SUPPLY

The demands on the Phase III power supply far exceed those of the Phase II satellites. In fact, the signals received by its users will be dependent largely on how much power can be produced by this system. For a given output requirement from the satellite, a greater power from the solar arrays can be used to increase the transmitter power and decrease the required antenna gain. This is desirable since it would relax the requirements on the attitude control system and allow the solar arrays to be oriented more favorably toward the sun. This in turn would produce more power. Unfortunately, the limitations on the power that can be generated are rather severe. Specifically:

1. The physical size of a secondary satellite (piggyback) is very restricted. Surfaces for mounting solar panels are limited.
2. Solar panels are very expensive. Present costs are about $3,000 per square foot.
3. Losses in distributing power within the spacecraft are about 10 percent.

Solar cells will produce 45 watts of power at the beginning of the spacecraft lifetime and will degrade to about 30 w output in five years. This much power will support a single 50 w PEP transponder which requires an average input power of 25 watts and will provide sufficient reserve to operate the remainder of the on-board systems throughout the satellite's lifetime. During eclipse periods, which can be rather severe in elliptical orbits of this type, transponder operation will be assured by using a nickel-cadmium battery similar to the type used on Phase II satellites. The battery will be able to recharge completely during the very long periods of sunlight near orbit apogee.

CONTINUING CONTROL VIA MICROPROCESSOR

With the advent of the microprocessor, minor design faults such as those that have become apparent from time to time in OSCAR 6 and 7 often can be corrected by remote control. In this way, the two OSCARs now in orbit are serving as orbiting laboratories for the designers of the Phase III.

With the untold benefits that will accrue from its ability to affect the control system after launch, a CMOS microprocessor will be utilized in a system to be known as the Integrated Housekeeping Unit. This system will consist of a command decoder, analog-to-digital converter, microprocessor, and at least 2048 bytes of random access memory (RAM).

Responsible for controlling virtually every function on board the spacecraft, it will execute all telemetry and command requirements, monitor the condition of the power and communications systems, and take corrective actions as necessary. It will establish clocks needed for various spacecraft timing functions, and it will interact with the attitude sensors and torquing magnet to adjust orientation of the spinning body in space, its most demanding task. In addition, the IHU will make the final decision on whether all on-board systems are "go" for the kick-motor firing. If confirmed, the computer, not a ground control station, will send the command to fire. The functional block diagram (Figure 3-8) summarizes the interconnection of Phase III components.

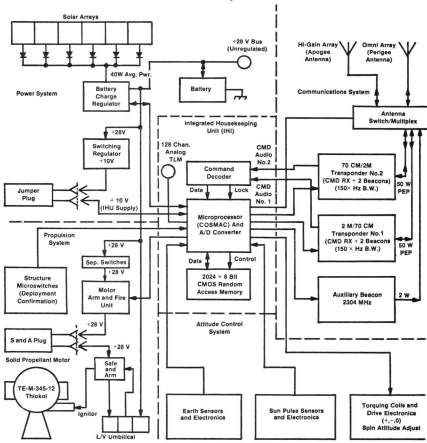

Figure 3-8 *Functional block diagram, AMSAT Phase III.*

THE AMATEUR/PROFESSIONAL

The special relationships between the professional and amateur worlds place a new set of constraints upon the amateur builder of space hardware. In addition, among the amateurs capable of designing and building amateur space hardware, volunteers are hard to find. Perhaps it's not so surprising. While the rewards are great when solving a challenging design problem that will benefit thousands of others, the pressures of the professional world are transferred to the shoulders of the amateur space volunteer.

Again, Phase III stretches this tenuous situation to the limit. Significant development effort is required, much of it not in electronics. Volunteers, like long-distance runners, must pace themselves. A steady work pace is required if the amateur space buffs who work on Phase III are to avoid "burn-out," or a sudden decrease in interest after a period of intense progress.

THE COST OF SATELLITES

How do you build these satellites for such a small amount of money? Part of the answer, as most amateurs already know, is that an incredible amount of free labor goes into such a project. But even if it were valued at $50 per hour, volunteer labor would not make up for the difference in cost between what amateur radio paid for AMSAT-OSCAR 7, and what the aerospace industry would have charged had they been asked to do the same job. The reasons for this difference in cost are many and complex. Simply stated, in thinking about designing and building a satellite of this sort, amateurs and professionals would approach the problem in completely different ways.

Most important, it is possible to construct reliable, long-lived satellites which will tremendously increase amateur capability at relatively low cost. How low is low? Commercial communications satellites typically cost between $10 and $50 million. A complete system, including ground stations, could easily top $100 million. In contrast, typical amateur installations range from a few hundred dollars to perhaps as much as $5,000. Amateur satellites have been, and can continue to be built and launched for $50,000 to $200,000—a range significantly below the cost of commercial systems, but far above the cost that amateurs must pay for their equipment. As a result, this new amateur/professional technology takes on a level of sophistication greater than amateur technology, but less than that of NASA technology. A new semiprofessional technology has been developed.

The Amateur Satellite Corporation has never, and does not intend in the future, to charge a fee for the communications provided by the satellites. There are terrestrial amateur repeaters which require that those who use them belong to a parent organization. AMSAT's policy, on the other hand, stresses that any form of membership requirement would limit the number of people using satellites. In addition, AMSAT believes that open satellite use by amateurs is the best way to encourage future satellite applications we cannot even dream of today. OSCAR 6 and OSCAR 7 have been linked together to extend the range of communication. Within the next decade, Phase III satellites will be linked together to provide reliable, round-the-world communication on a continuous 24 hour basis.

OSCAR IN THE CLASSROOM

Since the launch of OSCAR 6, this series of amateur satellite is becoming a popular educational aide. Students use OSCAR as a real-life example of the principles in physics, science, math, and engineering. Students in elementary schools, junior high, and high schools are listening to the slight frequency shift as OSCAR speeds through space. From this Doppler shift, the speed of the satellites can be determined. Students have been amazed to discover that their calculations show OSCAR tumbling in space as it proceeds through its orbit. The Talcott Mountain Science Center in Avon, Connecticut provides a curriculum supplement for educators who would like to set up an OSCAR station in their schools. By inserting the telegraphy data from OSCAR in specific formulas, students can calculate the life span of OSCAR's batteries and its general condition. The ARRL in Newington, Connecticut offers an OSCAR kit which tells you everything you want to know about OSCAR.

OTHER SATELLITE APPLICATIONS

Satellite technology is fast becoming a major part of the space race. Satellites are becoming indispensable "eyes and ears" of the Pentagon. They are used for picture taking, as heat-sensing spies, to communicate globally, and to navigate ships. The government uses them to watch global weather patterns and to monitor what crops are being grown throughout the world.

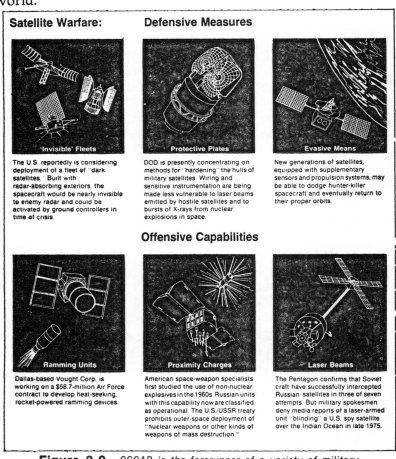

Figure 3-9 *OSCAR is the forerunner of a variety of military satellites now on the drawing boards.*

Courtesy of Machine Design.

The Air Force is tripling its budget for satellite and ground-control systems in 1979 to $37 million. Amateurs who develop satellite-related skills through OSCAR are getting in on the ground floor of this new and promising technological revolution. High ranking military officials who talk about future warfare being waged with satellites are extremely serious. Several military satellite programs are either on the drawing boards or in the design stage. Figure 3-9 summarizes both defensive and offensive military satellite techniques which you will be hearing much more about in the future.

4

A Practical Guide to Basic Ham Circuits: How They Work

HOW THE CRYSTAL SET BECAME THE TRF RECEIVER

In 1904, J.C. Bose in India discovered that a semiconductor named galena, if touched at the right spot with a fine steel wire (cat's whisker), made a sensitive detector of electromagnetic energy. Sending the energy of a powerful spark gap into a very long wire, Marconi was soon able to send audible telegraphy signals to a galena receiver across the Atlantic.

The publicity afforded this remarkable feat of wireless communication made it possible for enthusiasts the world over to begin building galena crystal receivers. The circuit in Figure 4-1 is an early crystal receiver. Table 4-1 gives an explanation of the symbols shown in Figure 4-1. The coil and capacitor (LC combination) is said to be tuned to only one frequency. This is the frequency where the capacitive reactance of $C(X_c)$ is equal to the inductive reactance of $L(X_L)$. The reader need not be awestricken by this language. Every coil has inductive reactance. This is how its coiled turns react to the flow of electrons, a form of resistance being created. Every capacitor has capacitive reactance. This is the way its capacitance reacts to the flow of electrons, a form of resistance being created.

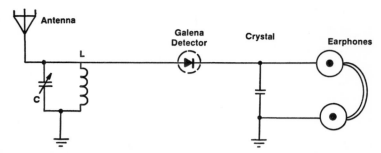

Figure 4-1 *The schematic diagram of the early crystal receiver. This circuit was the grandfather of all modern receiving circuits.*

When these two resistances, X_L and X_c, are equal, they are also opposite. That is, one pushes while the other pulls. The net effect is that electrons cannot pass through the circuit—they are trapped, or 180 degrees out of phase. When that occurs, no current flows; the impedance or AC resistance is said to be maximum. This condition occurs only at the tuned frequency or the resonant frequency. In Figure 4-1, it canbe seen that any current flowing from the antenna through LC to ground is consequently grounded out and will never reach the detector CR1. Thus the receiver is most sensitive only at the resonant frequency of LC where no current can flow to ground. In voice transmitters, we have a carrier which carries the radio signal over long distances. The voice is impressed upon this carrier, or is said to modulate this carrier. The carrier and its modulated audio signal encounters CR1 (galena crystal or germanium diode in Figure 4-1) and is rectified. This means that the carrier cannot pass through CR1 and is eliminated. However, the modulation or audio component will pass very easily.

When these receivers were used for telegraphy or CW, the transmitter was keyed on and off. When it was on, a hissing noise would be heard in the earphones corresponding to either a dot or a dash. The crystal receiver of Figure 4-1 proved to be satisfactory as long as the radio signal was extremely strong, or atmospheric conditions were extremely favorable. However, when the signal was not strong, it could not be heard—the receiver was not sensitive enough. Another problem was that the audio signal from the detector was too weak to drive a loudspeaker; only sensitive earphones could be used.

BASIC HAM CIRCUITS

SCHEMATIC SYMBOLS USED IN CIRCUIT DIAGRAMS

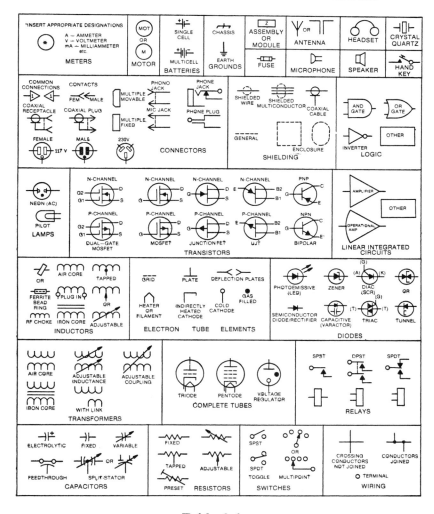

Table 4-1

Courtesy ARRL

Dr. Lee de Forest, called the father of radio, soon perfected the vacuum tube, or the audion, to solve these problems. The tuned radio frequency circuit of Figure 4-2 was developed. Tube V_1 was now placed ahead of the crystal detector to amplify the weak carrier signal before detection. A second tube, V_2, amplified the weak demodulated audio signal so that a loudspeaker might be used.

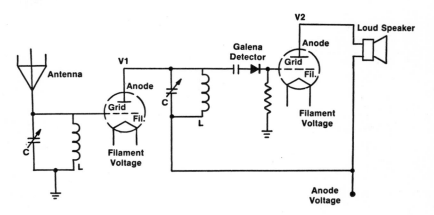

Figure 4-2 *The early crystal set has now grown into the Tuned Radio Frequency circuit. In vacuum tubes V1 and V2, a filament is heated. It then emits many electrons (negatively charged) to the positively charged anode. These electrons are intercepted by the grid which controls their flow. Thus a very small change on the grid controls a large anode voltage, and amplification results. The TRF receiver is thus far more sensitive than the early crystal receivers.*

THE SUPER-HET

The tuned radio frequency (TRF) receiver of Figure 4-2 was certainly an advance over the simple crystal set. However, more and more radio transmitters were coming on the air and the radio spectrum was becoming crowded. This TRF receiver had poor selectivity, that is, it could not discriminate between several radio signals whose frequencies were close to each other. This problem and the need for even greater sensitivity than the TRF receiver could offer led Major Armstrong, during World War I, to invent the super-heterodyne receiver. Major Armstrong found that the Q—quality factor, or maximum efficiency of a tuned LC circuit (remember the LC of Figure 4-1), was greatest at only one frequency. Not only did a higher Q give sharper tuning (greater selectivity), but it also gave less circuit loss, resulting in greater receiver sensitivity. However, the problem with the TRF receiver was that the LC circuits had to be tuned over many frequencies in order to receive more than one transmitting signal. How was it possible to receive over a broad spectrum, or band, of frequencies, when LC circuits have their optimum Q when tuned to a single frequency?

Major Armstrong devised a circuit called a local oscillator. Vacuum tube V_1 in Figure 4-3 is the local oscillator (LO). The LO is tuned in order to change the frequency which the super-heterodyne receiver is tuned to. The signal generated by the LO is mixed with the incoming radio frequency (RF) carrier we wish to receive. Mixing produces a beat frequency just as two musical notes played together produce a beat note which is the difference between those two musical notes. This beat frequency is fed to the IF (intermediate frequency) amplifier. The IF amplifier was the solution to the problem that Major Armstrong grappled with. Here we have two or more extremely high Q, LC circuits which are fixed tuned. That is, we tune the super-heterodyne by changing the LO frequency, but the IF frequency *always* remains the same.

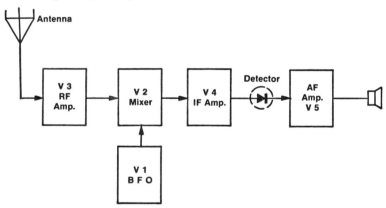

Figure 4-3 *A block diagram showing each state of the basic super-heterodyne receiver. The RF amplifier of the early TRF receiver has now become an IF amplifier (V4) with an extremely high gain. In addition to increased sensitivity, this receiver is also more selective than the old TRF.*

Let us illustrate this point. AM radios have an IF frequency of 455 kilohertz. If we tune the LO to 1455 kilohertz, the mixer will only receive 1,000 kilohertz (remember the difference frequency is 455 kilohertz). If we tune the LO down to 1255 kilohertz, then the mixer (V_2) will only receive 800 kilohertz, because the difference frequency is always constant. We can make the super-heterodyne even more sensitive and selective by adding a tuned radio frequency amplifier (TRF) stage (V_3) ahead of the mixer. The circuits following the IF amplifier (detector and V_5) are the same as would be found in the TRF receiver.

The term double-conversion super-heterodyne receiver is often heard when referring to modern receivers which operate in the VHF (very high frequency) and UHF (ultra high frequency) part of the radio or TV spectrum. Figure 4-4 is the block diagram of such a receiver.

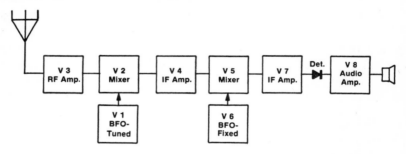

Figure 4-4 *Block diagram of the Double Conversion Super-heterodyne receiver. Mixing is accomplished twice and there are 2 IF amplifiers. Consequently, the selectivity (ability to tune out adjacent signals) is very sharp.*

In the double-conversion super-het, as the name implies, conversion, or mixing, occurs twice. V_2 feeds IF amplifier V_4 as is standard in the ordinary super-het. However, instead of V_4 output going to a detector, it is again mixed with a second LO, V_6, going then to a second IF amplifier, V_7. High Q, LC circuits in the IF amplifier (V_7) are tuned to the difference between the frequency of V_6 and V_4. Therefore, V_7 is tuned to 50 kilohertz. At this extremely low frequency, the Q of the LC circuits is extremely high, and their sharp bandpass characteristics, or tuning, result in extremely high receiver selectivity.

FM Receivers

We have studied the simple crystal receiver, the TRF, the super-heterodyne, and finally the sophisticated double-conversion super-heterodyne. Each of these have a common denominator. They all have a simple crystal diode, or vacuum tube diode detector. This detector discards the high frequency carrier, passing only the relatively slow variations in the carrier, otherwise known as the modulation. In Figure 4-5a, we see a fully modulated carrier. In Figure 4-5b, the carrier has been eliminated and only the demodulated audio variations remain. These variations are changes in the intensity, or amplitude, of the discarded carrier. That is why it is called amplitude modulation, or AM. In Figure 4-5c we see another modulated carrier, but this time the amplitude is constant. How then can information be impressed on the RF carrier? There always must be a variation that corresponds to the original information, whether it be audio, video, computer data, etc. A close examination of Figure 4-5c shows that the frequency itself (vertical lines) is changing. Indeed, this carrier has been frequency modulated at the transmitter. The diode detector we have studied responds to amplitude variations. In this case it would be useless.

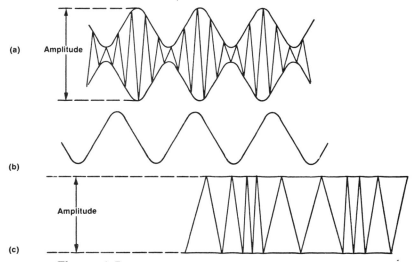

Figure 4-5 *In (a) we see the carrier with its modulated speech, or in television the modulation would be video. The carrier may be thought of as the horse which gallops across large distances, while the modulation may be thought of as the rider on the horse. In (b) the carrier is gone and only speech remains. In (c) we have another modulated carrier. This time the rider or modulation is changing the frequency, not the amplitude. Thus AM has now become FM.*

To demodulate or detect an FM carrier, we need a detector that will respond to variations in frequency, not amplitude. This type of detector is called a discriminator, or ratio, detector. A typical discriminator circuit is seen in Figure 4-6. We now have two diodes (CR1, CR2) instead of one. The voltage across the series load resistors R1 and R2 is equal to the algebraic sum of the individual output voltage of each diode. A signal at the IF midfrequency will produce equal and opposite (plus 5^v, minus 5^v) voltages, with the output voltage being zero. As the RF signal varies from this midfrequency, the individual voltages become more unequal, producing a larger audio voltage at the output. Each half of the secondary goes to ground through the series load resistors. At midfrequency, the RF voltage across the secondary is 90 degrees out of phase with that across the primary. Since each diode is connected across one-half of the secondary, and the primary is in series, the resulting RF across each is equal and opposite, or zero output. When the frequency varies, a voltage proportional to the difference between the RF voltage across the two diodes exists across R1 and R2. As the signal frequency varies back and forth across the resonant frequency, an AC voltage of the same frequency as the original modulation is produced.

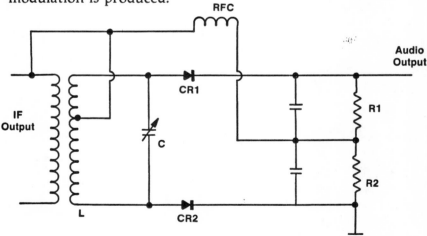

Figure 4-6 *Circuit of an FM discriminator. The old AM crystal detector has become much more sophisticated. The additional crystal and the split secondary winding of L enables FM to be detected.*

BASIC HAM CIRCUITS

In addition to the FM detector, we can add a limiter circuit before the detector. The AM receiver responds to amplitude variations, therefore we can remove some annoying static (an amplitude variation) but not too much, for we would then lose the audio information we wish to recover. However, in the case of FM there are no amplitude variations to begin with, thus any amplitude variation is static and it can be removed. A limiter circuit simply chops off any amplitude variation found in the FM carrier. This explains static-free FM reception.

Receiving Morse Code and Single Sideband

In the super-heterodyne circuit, we discovered that a beat frequency oscillator (BFO) generated a signal that was mixed with the antenna signal to produce the IF frequency. The IF frequency is too high to be audible. It may be 50 kilohertz, 455 kilohertz, 4.5 megahertz, or 10.5 megahertz. In each case, these frequencies are too high to be perceived by the human ear. The human ear can hear a tone only up to about 15 kilohertz. If we added a BFO to a receiver whose frequency was 2 kilohertz away from the IF frequency, then it would produce an audible beat note with any signal coming from the antenna. What purpose could be served by producing the beat note whenever the antenna picked up a signal?

Remember how Marconi sent electromagnetic waves across the Atlantic to a crystal set? A dull hissing noise was heard in the earphones each time the transmitter key was closed on the other side of the Atlantic. Whenever the radio signal was weak, some "hissing" dots or dashes were lost. If the telegrapher wanted to send Morse code faster, the hissing dots were even more easily lost. If those hissing dots and dashes could be changed into a smoother, high pitched tone, then the fastest and the weakest code signals could be heard clearly and distinctly. The BFO separated from the IF frequency by 2 kilohertz does exactly this. It beats with the incoming telegraph signal to produce a smooth, high pitched whistle or tone.

Single Sideband

This same BFO has yet another function. While AM modulation and FM modulation vary either the amplitude or the frequency of a carrier, there is yet one more form of modulation—single sideband. An SSB transmitter will be covered later in this chapter. Essentially, this type of transmitter sends only a high frequency sideband (audio mixed with carrier) adjacent to the carrier itself. The carrier itself, however, has already been eliminated. Thus, the lost carrier must be reconstructed again in the receiver. The receiver is tuned until the received sideband is 2 kilohertz away from the BFO. The receiver BFO then is turned on and becomes the lost SSB carrier.

TRANSMITTERS

The Spark Gap Transmitter

Going back to 1906, Marconi fed the energy from a powerful spark into a long wire antenna. The electromagnetic signal that was subsequently sent over the airwaves had an extremely broad spectrum. That is to say, its energy content was spread over many thousands of cycles. The spark-gap transmitter operated in conjunction with receivers that were very broad also. Major problems developed when more transmitters came on the air; they soon began interfering with each other.

When the Audion, or vacuum tube, was invented, it was found that it could generate radio frequency energy. Figure 4-7a shows how the vacuum tube was made into a generator. Remember that the grid is a control valve that regulates the flow of a large number of electrons from cathode to anode. Part of the amplified electron variation from the anode is fed back to

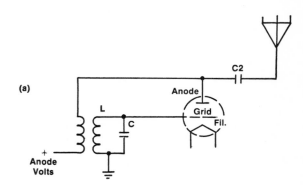

BASIC HAM CIRCUITS 85

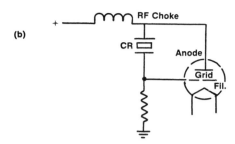

Figure 4-7 In (a) a coil winding at L feeds energy back to the grid. Because the feedback is inductive (through a coil) this is called a Hartley oscillator. If the feedback path were capacitive (through capacitors) then it would be called a Colpitts oscillator. In (b) the energy feedback path is now through a quartz crystal. This is the Pierce crystal oscillator.

the grid again, reinforcing the grid variations. This process becomes a self-sustaining one, and the tube is said to be in oscillation. In other words, it continues to generate a high frequency signal whose frequency depends on the resonance of LC.

If we attach C_2 to a very long wire, then we can broadcast this electromagnetic energy out over the air waves for a considerable distance. However, temperature changes and other factors affect LC, changing its resonance. This changes our broadcast frequency and creates many problems with highly selective receivers. In Figure 4-7b, we have replaced the LC feedback circuit with a quartz crystal feedback circuit. A slab of quartz will have an electromechanical or piezoelectric resonance or vibration according to the precise dimensions it is cut to. In the crystal oscillator, feedback can occur only at the crystal resonant frequency, hence RF energy is only generated at this one frequency. Thus the crystal oscillator is extremely stable. We have solved one problem only to be faced with another. The quartz crystal is extremely fragile. If too much power from the anode is impressed upon the crystal (CR), it will fracture. Thus the oscillator must always generate very low power. As we need higher power to radiate our radio wave over a considerable distance, we must add additional stages. In Figure 4-8, we have tacked on a buffer and power amplifier.

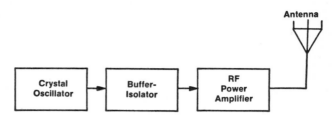

Figure 4-8 *Block diagram of a complete high power telegraph transmitter. The delicate oscillator cannot deliver too much power so 2 more stages have been added.*

The buffer isolates the sensitive oscillator from the high power amplifier. This amplifier can then provide thousands of watts of power to the transmitting antenna. To calculate the power input of the transmitter in Figure 4-8, multiply the voltage on the power amplifier anode by the current of the power amplifier. For example, if the power amplifier anode voltage is 1000 volts, and its current is 0.4 amps, then the transmitter power input is 400 watts. If we turn this transmitter on and off with a telegraph key, we can send Morse code across the Atlantic as Marconi did. In addition, thousands of these stable transmitters can operate on different frequencies simultaneously, and they will not interfere with each other. That is certainly an advancement over Marconi's crude spark-gap transmitter.

After the invention of the vacuum tube, radio amateurs looked for a way to send their voices over the air waves. They knew that transformers could transfer the human voice from one telephone circuit to another. So, hams added modulation transformers and audio amplifiers to their telegraph transmitters. The AM radio telephone transmitter of Figure 4-9 was the result. The pre-amp, speech amplifier, and audio power amplifier simply boosted the audio power so that it could modulate the RF carrier 100 percent (Figure 4-5a). The audio power amplifier must supply 50 percent of the carrier input power to fully modulate that carrier. The rule of thumb is to calculate that an audio power amplifier is only 50 percent efficient; 200 watts input (voltage on anode times current) will yield only 100 watts of actual audio power. If the RF power amplifier input (anode volts times current) is 200 watts, then

BASIC HAM CIRCUITS

one-half audio power, or 200 watts of audio at 50 percent efficiency, is required for 100 percent modulation.

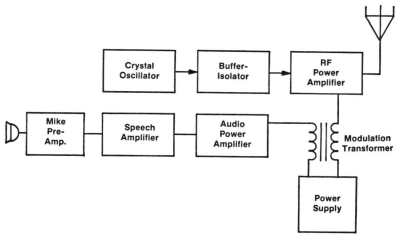

Figure 4-9 *Several more stages have been added to the telegraph transmitter so that the carrier can be modulated. Thus the telegraph transmitter has now become the telephone transmitter.*

FM Transmitters

As we have seen in the AM radiotelephone transmitter, one-half the total power consumption (400 watts) powers the RF carrier (200 watts), and the other half provides audio modulation. As amateurs began to build mobile transmitters to operate from their autos, they soon discovered that the AM phone transmitter (with its large audio power requirement) created a large drain on their auto battery. In addition, AM radio was subject to constant static and the weak transmissions from mobile transmitters were seldom received well. Hams soon found a solution to both of these problems—FM. Using FM, all the amplitude variations, including static, could be eliminated by the receiver's limiter.

Equally important, the audio power amplifier with its heavy power consumption could be eliminated, thus reducing wear and tear on the auto battery. How is this possible? Remember, FM modulation does not vary the amplitude of the carrier, but does vary its frequency. Instead of all the audio power required in Figure 4-9, a reactance modulator (Figure

4-10) is placed across the oscillator crystal. This modulator does not provide power but rather a change in capacitive reactance (X_c), which causes a very slight shift in the crystal's resonant frequency. This minute shift is then multiplied as much as 32 times to increase the frequency shift at the transmitter's output—called frequency deviation. In Figure 4-10, the resonant frequency of the crystal is 6.875 megahertz. The chain of frequency multipliers doubles and triples this frequency 32 times until the carrier output at 220 megahertz is achieved. The speech amplifier shifts the crystal frequency ±250 hertz. After multiplication 32 times (250×32), the final frequency deviation is 8,000 hertz. That is, audio modulation will shift the RF carrier plus and minus 8,000 hertz for 100 percent modulation. Amateur FM transmitters typically produce a deviation from 5 kilohertz to 15 kilohertz, while FM broadcast stations have a larger share of the spectrum (88-108 megahertz) and can produce a frequency deviation up to 75 kilohertz.

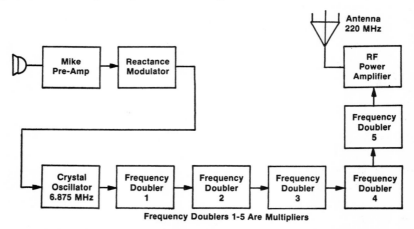

Figure 4-10 *Block diagram of an FM transmitter. The reactance modulator creates a small frequency change in the 6.875 MHz crystal oscillator. Frequency multipliers merely raise the oscillator frequency to the desired frequency which is then fed into the antenna.*

Single Sideband Transmitters

SSB is a form of AM with several advantages over AM. In AM, a 200 watt transmitter consumes 200 watts for the carrier

BASIC HAM CIRCUITS

and 200 watts of audio, or 400 watts. In SSB, a 200 watt transmitter only consumes 200 watts during full speech power or talk power, 100 watts during one-half talk power, and less when there is no speech at all. What is a sideband? When an AM transmitter is modulated, the audio (50 hz – 10 kHz) is mixed with the carrier it modulates. Whenever several frequencies are mixed as in the super-het receiver, additional frequencies are developed. In AM modulation, these additional frequencies slightly above and below the RF carrier are called sidebands, or audio-RF products. A receiver's AM detector converts both sidebands back into audio. In SSB, a filter removes one of these sidebands at the transmitter. Another device, a ring modulator or balanced modulator, cancels out the carrier itself. Thus, only one sideband remains and is transmitted over the air.

One advantage of SSB is that the absence of carrier eliminates the annoying whistle, or beat-note, when two transmitters are operating at a very close frequency. Thus the amateur spectrum can accommodate many more SSB stations than either AM or FM. Another advantage is that SSB talk power is instantaneous, not continuous as in AM. Tubes and other components which could only handle 100 watts continuously (AM), can now handle 400 watts instantaneously. Thus the 'effective' power when using SSB can be much higher.

5

How to Understand Antennas

BASIC THEORY

In the last chapter, we learned how early amateur gear was merely a spark gap and an antenna, or a galena detector and an antenna. Even long distance communication could take place with this crude equipment if the antennas were efficient enough. A general knowledge of basic antenna theory will guide us in making sure that our antenna system is efficient. What determines antenna efficiency? The key factors of antenna efficiency are antenna current, directivity, bandwidth, radiation angle, and polarization.

ANTENNA CURRENT

Our transmitter feeds its RF energy through a transmission line into the antenna. We hope that the antenna will radiate the strongest possible signal into the ether. Another way of saying this is that it will produce the strongest possible field strength. The field strength is directly proportional to the current flowing in the antenna. This current becomes electrostatic energy traveling through the atmosphere. If you throw a small pebble (small current) into the water, a small ripple will reach out over a short distance. If you throw a large stone (high current) into the water, a large ripple will reach out over a long distance. How do we get the largest possible current into the antenna? The antenna must be tuned to the frequency of the wave that is being fed into it. Also, the transmission line must be properly matched to the antenna.

Remember in Chapter 1 the wavelength of a particular frequency was discussed. A 30 megahertz signal has a wavelength of 10 meters. All antennas are based on some variation of a half wave of the transmitting wavelength. In this case, 18 feet. A half-wave antenna for 10 meters would then be 18 feet long. However, the actual antenna will be slightly shorter than the half wave in free space. This is so because large diameter antenna rods tend to lengthen the antenna, so it must be shortened to compensate. The insulators at both ends of a wire antenna also tend to lengthen the antenna, so it must be shortened to compensate. The chart in Figure 5-1 gives the lengths of half-wave wire antennas for the popular amateur bands. These lengths are based on the formula:

$$\text{length of half-wave antenna} = \frac{468}{\text{frequency in megahertz}}$$

The Half-Wave Antenna

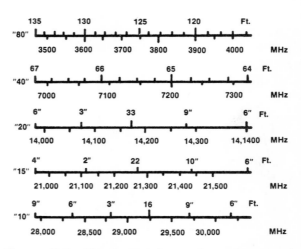

Figure 5-1 *The above scales can be used to determine the length of a half-wave antenna of wire.*

Why was the half wave chosen for the most efficient antenna? Remember the sine wave? It started at zero voltage,

HOW TO UNDERSTAND ANTENNAS

went to maximum voltage, to zero, to maximum again, and back to zero. Figure 5-2 shows the radio sine wave as it travels in space. If we cut a wire whose length is one-half wave, from A to B, its ends will correspond to two successive high voltage points of the wave. As both ends are open (connected only to insulators), they are said to be voltage points, favoring voltage points of the transmitter's 30 MHz radio wave. In other words, this wire is now turned to 30 MHz.

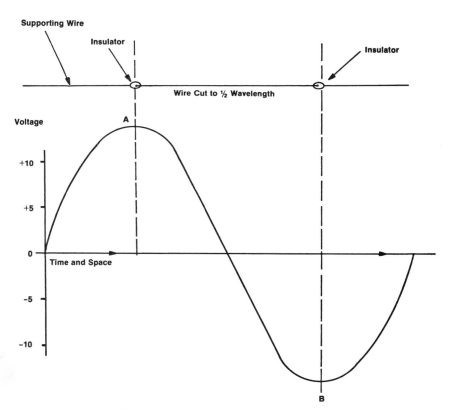

Figure 5-2 *Voltage variation on a wire a half wavelength long. The connection to the transmitter has been omitted.*

DISTRIBUTION OF VOLTAGE AND CURRENT

Any half-wave antenna is composed of capacitive and inductive reactance; X_c and X_L. Both X_L and X_c slow down the voltage or the current. This is why the rising and falling (AC) current on an antenna occurs at exactly the opposite place as the rising and falling (AC) voltage. Figure 5-3 shows the distribution of voltage and current on an antenna wire a half wave long. At the voltage or open ends, the voltage is maximum while the current is minimum. Current is maximum at a current loop and minimum at a current node. The important thing to remember about the current and voltage distribution is the impedance.

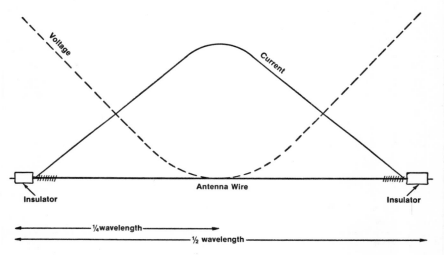

Figure 5-3 *The distribution of voltage and current along the half wavelength wire. Current is maximum at its center while voltage is maximum at its ends.*

IMPEDANCE

To understand impedance (Z) it is necessary to understand resistance (R) and reactance (x). Resistance is just what it sounds like—opposition to the flow of electrons. A very thin wire, a resistor, or any resistive component can offer resistance to a flow of current. Resistance is measured in ohms, kilo-ohms, or meg-ohms (1000 and 1 million ohms). Resistance is a satisfactory unit of measurement for DC (direct current) circuits. However, in an AC (alternating current) circuit the opposition (or reinforcement) of current becomes more complex. In AC circuits, the status of current and voltage is measured as impedance.

In AC circuits, reactance is at work as well as resistance. Reactance can be either capacitive reactance (X_c) or inductive reactance (X_L). It takes time for voltage to build up while a capacitor is charging, thus the voltage is delayed in capacitive reactance. It takes time for current to build up when an inductor is placed in an AC circuit, thus the current is delayed in inductive reactance. AC circuits are usually a combination of X_c and X_L. Thus there is always a complex time (PHASE) relationship. This phase relationship determines the reactance. If two equal voltages were 180 degrees out of phase, they would cancel each other. Then there would be zero volts. However, if two equal voltages were perfectly in phase, they would add to each other. Then the total voltage would be double. The electric eel illustrates the phase principle. The electric eel has numerous cells, each producing 1.8 volts. When 600 such cells function together, then the eel can produce 1,080 volts. In the same way, small voltages in phase with each other can produce a much higher voltage.

The important thing to remember is that a high voltage, low current AC circuit has high impedance while a low voltage, high current AC circuit has low impedance. The resistance (R) and the reactance (x) determine the ratio of current to voltage. A large ratio means low impedance; a small ratio means high impedance.

FEEDING ENERGY TO THE ANTENNA

All transmission lines which feed energy from a transmitter to an antenna have a characteristic impedance. Impedance might be compared to an AC resistance. If current is low and voltage is higher along a transmission line, it is said to have a higher impedance. Or, if current is high and voltage is lower along a transmission line, it is said to have a lower impedance. The spacing between the two wires (or the center conductor and the shield in coaxial cable) and the diameter of the two wires determine the characteristic impedance of the transmission line.

Figure 5-4 lists the various types of transmission lines, their characteristic impedance (Zo), and the attenuation (loss) per 100 feet. For example, at 50 MHz (6 meters), 3.1 DB of power is lost when using 100 feet of RG58/A-AU coaxial cable. A 3 DB loss is equivalent to cutting your power in half. In other words, if your transmitter sent 40 watts into this coaxial cable, 100 feet away, only 20 watts would be available at the antenna. You may very well ask, why not always use the coaxial cable with the lowest loss? The answer is cost. The highest quality cable might cost 20 cents a foot. If it were necessary to run 200 feet, you might not wish to spend $40 for the transmission line alone. Figure 5-4 indicates that the characteristic impedance of transmission line can vary between 50 ohms and 450 ohms.

Transmission-Line Losses

Characteristics of Commonly-Used Transmission Lines

Type of Line	Z_0 Ohms	Vel. %	pF per ft.	OD	Attenuation in dB per 100 feet						
					3.5	7	14	21	28	50	144	420
RG58/A-AU	53	66	28.5	0.195	0.68	1.0	1.5	1.9	2.2	3.1	5.7	10.4
RG58 Foam Diel.	50	79	25.4	0.195	0.52	0.8	1.1	1.4	1.7	2.2	4.1	7.1
RG59/A-AU	73	66	21.0	0.242	0.64	0.90	1.3	1.6	1.8	2.4	4.2	7.2
RG59 Foam Diel.	75	79	16.9	0.242	0.48	0.70	1.0	1.2	1.4	2.0	3.4	6.1
RG8/A-AU	52	66	29.5	0.405	0.30	0.45	0.66	0.83	0.98	1.35	2.5	4.8
RG8 Foam Diel.	50	80	25.4	0.405	0.27	0.44	0.62	0.76	0.90	1.2	2.2	3.9
RG11/A-AU	75	66	20.5	0.405	0.38	0.55	0.80	0.98	1.15	1.55	2.8	4.9
Aluminum Jacket, Foam Diel.[1]												
3/8 inch	50	81	25.0	–	–	–	–	0.36	0.48	0.54	0.75	1.3	2.5
1/2 inch	50	81	25.0	–	–	–	·	0.27	0.35	0.40	0.55	1.0	1.8
3/8 inch	75	81	16.7	–	–	–	–	0.43	0.51	0.60	0.80	1.4	2.6
1/2 inch	75	81	16.7	–	–	–	–	0.34	0.40	0.48	0.60	1.2	1.9
Open-wire[2]	–	97	–		0.03	0.05	0.07	0.08	0.10	0.13	0.25	–
300-ohm Twin-lead	300	82	5.8		0.18	0.28	0.41	0.52	0.60	0.85	1.55	2.8
300-ohm tubular	300	80	4.6		0.07	0.25	0.39	0.48	0.53	0.75	1.3	1.9
Open-wire, TV type												
1/2 inch	400	95			0.028	0.05	0.09	0.13	0.17	0.30	0.75	–
1 inch	450	95			0.028	0.05	0.09	0.13	0.17	0.30	0.75	–

[1] Polyfoam dielectric type line information courtesy of Times Wire and Cable Co.
[2] Attenuation of open-wire line based on No. 12 conductors, neglecting radiation.

Figure 5-4 *Characteristics of commonly used transmission lines.*

Courtesy ARRL

In the previous section, it was mentioned that the impedance of the antenna was the important thing to remember. This means that the chartacteristic impedance of the transmission must equal the impedance of the antenna where the transmission line is connected to the antenna. That is the only way the antenna will accept all the energy available from the transmission line. For example, if we use RG58/A-AU, this cable must be connected to the antenna where the antenna's impedance is 53 ohms. This is a very low impedance and corresponds to the point on the antenna where the current is highest and the voltage lowest. Thus the center of the half-wave antenna would be the most appropriate place to connect a transmission line with a low impedance (see Figure 5-3).

If we mistakenly connect an open-wire, one inch, transmission line to the center of the half-wave antenna, almost no energy will flow from the transmission line into the antenna. This type of transmission line has a characteristic impedance of 450 ohms. We are then matching 450 ohms to the center of the half wave, where the antenna's impedance is about 50 ohms. This would be called a mismatch of over 8 to 1, and almost no current would flow in the antenna. Instead, the energy (with nowhere to go) would bounce back and forth along the transmission line. This would create large standing waves.

While it is normal to have a rising and falling voltage and current on an antenna, this condition should not exist along the transmission line. When the transmission line is matched perfectly to its antenna, there are no standing waves; the line is said to be flat. That is, there are no voltage variations, thus the ratio of voltage to current is the same along the entire transmission line. The ratio is then one to one. This is called the transmission line's voltage standing wave ratio, or VSWR. The perfect VSWR is one to one, when there is a perfect impedance match. As more energy from the transmission line is reflected back onto itself, the VSWR becomes 2:1, 4:1, 6:1, or progressively worse. VSWR meters (now inexpensive) are placed between the transmission line and the transmitter. They indicate how much of the available energy is going into the antenna and how much is being reflected back along the transmission line.

VERTICAL ANTENNAS

In the section on the half-wave antenna, we learned how it can be center fed with a low impedance transmission line.

Remember, high current means low impedance or low resistance to current flow. Low current means higher impedance or more resistance to current flow.

Having fed the half wave at its center point, we have divided the half wave into two separate quarter-wave sections, one on each side of the transmission line. The horizontal half-wave antenna possesses several disadvantages. Its angle of radiation is relatively high. That is, much of its radiation is relatively high and leaves the earth at a high angle going into the sky. When you wish to use ground wave with nearby stations or talk to stations very far away, a low angle of radiation (energy hugs the ground) is desirable. The half wave must be twice as long as the quarter wave. A half-wave antenna on the 80 meter amateur band must be 120 feet long.

The vertical antenna overcomes these problems of the half-wave horizontal. The vertical quarter wave uses the ground or ground radials as the other quarter wave. Radio broadcast antennas use ground wave and thus are always quarter-wave vertical antennas working against ground. In addition, the vertical quarter wave has a low angle of radiation in all directions. Thus it is excellent for ground wave or working very distant stations. The vertical quarter wave is fed at its base with very low impedance transmission line.

LOADING COILS

Apartment dwellers, those who wish to use the low frequency bands from an auto, and those who want antennas that can be quickly erected and dismantled, all have something in common. They must find some way to make a shorter antenna do almost the same work as a longer one.

For example, to operate 80 meters from a car, a quarter-wave antenna would be 60 feet long. Obviously, this would be impossible unless we wished to suspend it from a helium balloon 60 feet up. The 60 foot length represents a given inductance (more wire means more inductance) spread out in space. Instead of having a straight wire, it is possible to use a coil which represents the same amount of inductance as a straight wire in space. Thus our 60 foot, 80 meter antenna now becomes an 8 foot whip attached to a loading coil at its base. Electrically,

HOW TO UNDERSTAND ANTENNAS

this arrangement is equal to the 60 foot quarter-wave antenna, but the 8 foot antenna will not be as efficient as the 60 foot antenna. However, being tuned to 80 meters and having its base impedance low, it will accept all the power from the transmission line.

In the above instance, the antenna has been shortened drastically. A trade-off was made between efficiency and convenience. It is also possible to reduce the antenna's length only slightly, with a slight reduction in efficiency. A quarter-wave CB antenna must be 9 feet long. A small loading coil (with a few turns) will permit us to reduce the required length to 5 feet. In this instance, the loss in efficiency will be relatively small. Figure 5-5 illustrates a typical mobile antenna with loading coil.

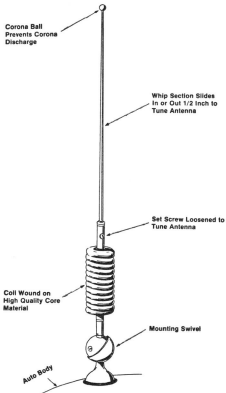

Figure 5-5 *Typical mobile antenna with loading coil. This antenna is base loaded.*

POLARIZATION

A vertical antenna is vertically polarized, a horizontal antenna is horizontally polarized. Polarization indicates the plane of the radiated wave relative to ground. The important fact to remember about polarization is to avoid cross-polarization. If the transmitter's antenna is vertical, the receiving antenna should be vertical; also, if one antenna is horizontal, the other should be horizontal.

DIRECTIVITY

The angle of radiation of an antenna actually refers to the direction of transmission in the vertical plane. Figure 5-6 shows an antenna system with various angles of radiation. When we use the word directivity, we refer to the direction of transmission (or favoring reception) in the horizontal plane.

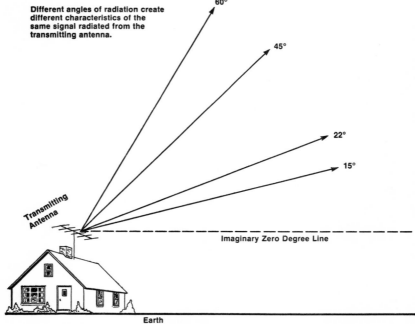

Figure 5-6 *Transmitting antenna radiated at a very high angle, 60°. Energy radiated at 60° will bounce back to earth not too far away. Energy radiated at 15° will bounce back to earth far away; it is also useful for ground wave activities.*

U.S. amateurs who wish to talk to Europe would certainly not want any of the signal to be sent in any direction except due east. The half-wave doublet or dipole will favor east and west only very slightly, as can be seen in Figure 5-7. In Figure 5-8, a director and reflector has been added to the dipole. We now have a three-element beam. With its director broadside to

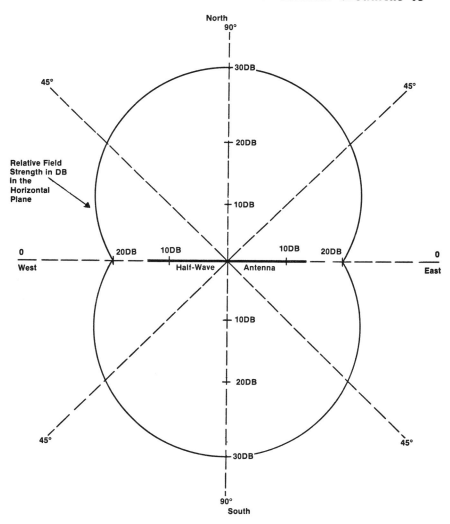

Figure 5-7 *Half-wave doublet has only a slightly directional radiation pattern broadside to the antenna.*

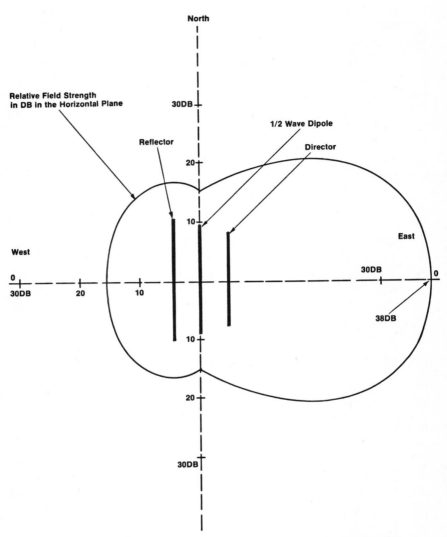

Figure 5-8 *Half-wave doublet is now turned north and south. A director and reflector are added to the antenna.*

Europe, we can have a forward gain in excess of 8 decibels. What has been gained? Every 3 decibels (DB) of gain is equivalent to doubling the transmitter's power. A 100 watt transmitter attached to a directive beam antenna with a 3 DB gain would be the equivalent of a 200 watt transmitter in its most favored direction: 400 watts would have a 6 DB gain; 800 watts a 9 DB gain; and 1600 watts a 12 DB gain. Thus it is more advisable to devise sharply directive antenna systems for point-to-point communication than to raise the actual transmitting power.

THE YAGI

Should we desire to increase directivity even more than that obtainable with the three-element beam of Figure 5-8, we can add more directors and possibly another reflector. A Yagi can have from three to nine elements. Doubling the number of elements gives a 3 DB gain. In other words, if a three-element beam has an 8 DB forward gain, then a six-element Yagi has an 11 DB gain. The greater the number of elements, the more forward gain is possible. However, the limiting factors are the size of the Yagi, its wind resistance, the size of the rotator needed, and the strength of the supporting mast or tower. Another limiting factor is the ability to accurately point the Yagi. For example, the nine-element Yagi (with a 14 DB gain) illustrated in Figure 5-9 would have an extremely sharp directive pattern. Even if it were pointed 30 degrees away from the target, much of its value would be minimized.

Yagis are used above 50 megacycles (6 meters) where the length of a half wave is relatively small. They would certainly be prohibitive on 40 meters, where the length of a half wave is 60 feet. However, two- and three-element beams are now in use on these lower frequency amateur bands. This is made possible by the use of loading coils in each element. A 40-meter dipole, or driven element, would normally be 60 feet. With two loading coils, it can be reduced to 20 feet. In Chapter 7, an eleven-element Yagi is described, which you can build yourself. Its dimensions are under 5 feet long and 1 foot wide, and it will produce a gain of approximately 18 DB on 432 MHz. It can be bolted to the top of a walkie-talkie and beamed to orbiting satellites by pointing your handheld walkie-talkie. The 18 DB forward gain will give your 5 watt walkie-talkie the equivalent power of a 320 watt transmitter. This will enable you to use OSCAR 7 and OSCAR 8 satellites (also the future Phase III satellites) when they are in the farthest reaches of their orbit.

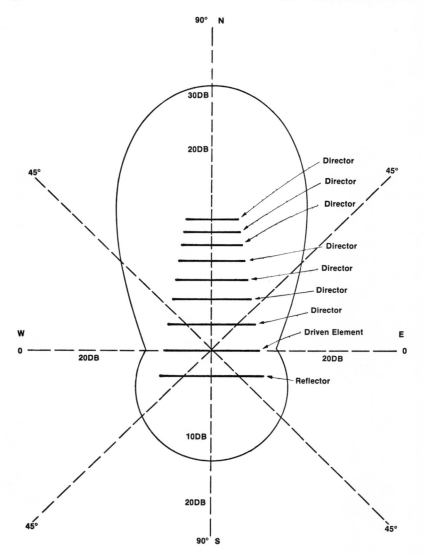

Figure 5-9 *This 9 elment Yagi has a 14 DB gain in the forward direction—north. Very little radiation exists off the back or the sides.*

HELICAL WHIP

We have seen how a loading coil placed at the base of an antenna increases its inductance and therefore its resonant frequency. We might call the loading coil a lumped inductance because it is lumped together. The coil may be at the base of the antenna (base-loaded) or in the center of the antenna (center-loaded). One disadvantage of a loading coil is that it makes the antenna highly selective. That is, the antenna will only load up or take RF at the tuned frequency. If the energy is fed to the antenna even slightly away from its resonant frequency, the VSWR goes up and the antenna will not load up.

Another type of short antenna that is not as sharply tuned as the loading coil antenna is the helical whip. Instead of having a loading coil, almost the entire antenna is wound on an insulated rod with a small diameter. Figure 5-10 shows the

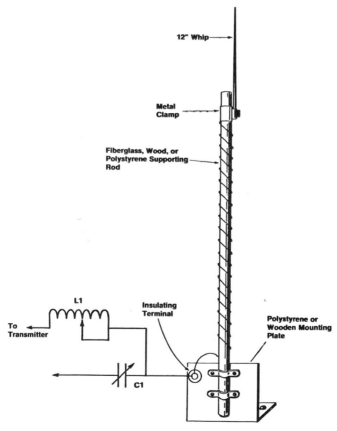

Figure 5-10 *The helically wound short vertical antenna. L1 lengthens the antenna; C1 shortens the antenna.*

helically-wound short antenna. The insulated supporting rod can be fiberglass, a treated wood dowl, or polystyrene. The helical is a quarter-wave vertical working against the earth as a ground. The supporting rod may be anywhere from 4 to 25 feet long. The longer the rod, the closer it is to an actual quarter-wavelength, the better the antenna's performance. Wooden stock may be coated with fiberglass or several coats of exterior spar varnish. This will improve the dielectric qualities of the wooden pole. Spar varnish can then be used again after the coil is wound to weatherproof the antenna. The helical coil may be No. 12 or 14 insulated copper wire. The spacing between turns should be equal. A 12 inch whip can be mounted on top of the supporting rod, as shown in the illustration. Two U clamps attached to a base supporting plate are capable of supporting this antenna. A multisection 365 mmfd capacitor (with sections in parallel) can be tuned to effectively shorten this antenna, or a small variable inductor (with a high Q) can be tuned to effectively lengthen the antenna. Tuning, however, will be far less critical than the same antenna would be with a loading coil configuration.

THE CUBICAL QUAD

The Quad is an antenna often used for long distance contacts where limited space exists. It has a driven element composed of four quarter-wavelength wires folded into a box-like shape. The entire loop is then a full-wavelength. Figure 5-11 shows the Quad. Behind the driven element is another full-wavelength loop. This loop is the reflector and has a tuning stub which is adjusted for maximum forward gain or directivity. The entire antenna system can be supported by a fiberglass cross-arm assembly or bamboo poles. Needless to say, this beam antenna cannot be rotated and is only useful for directivity in a given direction. The Quad has a low angle of radiation in the vertical plane, making it efficient for long distance transmission.

INVERTED V

A form of half-wave antenna that has come to be popular in suburban and urban areas is the "inverted V." Usually, two

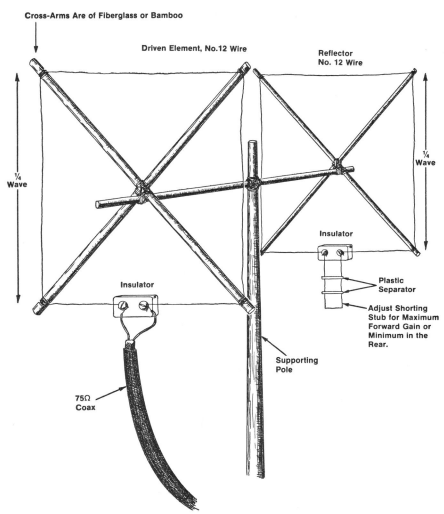

Figure 5-11 *The cubical quad with the tuned reflector. Directivity is away from the reflector. The shorting stub is adjusted for minimum field strength behind the reflector.*

quarter-wavelengths of wire are stretched between two poles or trees. When the lack of space prohibits this, then a single pole supports the center of the half-wave doublet. The two quarter-wavelengths of wire are then brought to ground with a high quality insulator near the earth where the two wires are anchored. Figure 5-12 shows the inverted V. It is a good

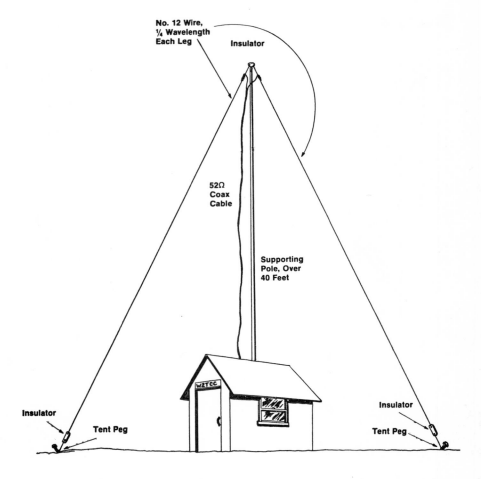

Figure 5-12 *The inverted V antenna, used where space is at a premium, but the full half-wave length is desired.*

antenna for long distance work if the center pole is at least 50 feet high. One disadvantage of the inverted V is that the center pole must be metal in order to be strong. This metal pole drops below that part of the antenna (the center) that is radiating the most energy (highest current). Thus the presence of the grounded metal pole absorbs some of the RF energy being radiated by this type of antenna. The wooden mast described later in this chapter might make an excellent supporting pole for the inverted V.

COLLINEAR AND PARABOLIC ARRAYS

At very high frequencies, a half-wavelength in space is small, and aluminum rods are used as elements. We have seen how many elements can be combined to form a Yagi. A collinear array consists of a second Yagi in the same plane and is placed a

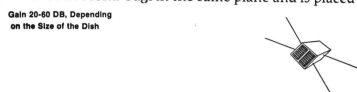

Figure 5-13 *The parabolic dish antenna. The dipole is in the center of the metal reflector, which is very large in terms of wave lengths.*

wavelength below (or above) the original Yagi. A collinear may also be as many as four vertical Yagis stacked together with as many as 128 elements. At very high frequencies, these arrays have fantastic directivity. However, tuning them is extremely critical. Their directivity is so sharp that if they are pointed only slightly away from their target, their value will be negated.

Another highly directive UHF antenna is one that we often see in science fiction movies. It is the parabolic dish antenna. Figure 5-13 shows the tiny 70 centimeter dipole (13 inches long) at the center of the dish. The metal dish is not grounded. Its surface resembles many, many reflectors to the short 70 centimeter wave, just as a flashlight's reflector resembles thousands of half-wave reflectors to the visible light coming from the flashlight bulb. Parabolic dish antennas used by NASA can be rotated according to azimuth and elevation by computers. These computers calculate the precise position of a satellite, missile, or other target. Inexpensive microcomputers are now available that track amateur Yagis so they can follow OSCAR, the amateur satellite.

MASTS

If a 40 foot high mast is required, wooden 2 x 3s or 2 x 4s may be used to construct the support. A single 22 foot 2 x 2 can form the top section. This is bolted to two bottom sections (each 22 feet long) which form a narrow triangle, their base spread to about 6 feet apart. An arrangement consisting of a 20 foot 2 x 4 bolted to two supporting 2 x 4s can give 40 feet of height. Another section bolted to this consisting of four 2 x 4s would give the entire mast 60 feet of height.

Electronic distributors now sell both aluminum and stainless steel poles of 10 feet in length for $5 each. One can be inserted in the other. With one set of guy wires 30 feet up, and the other set at the top, an antenna support 60 feet high can be erected. For heights 60 feet and above, a collapsible tower is recommended. These fold-over towers can be lowered to examine antennas and make adjustments. If you use the top of a tree as a supporting mast, make sure to use two strong springs at each end. There is another way of preventing the antenna from snapping when trees bend in the wind. Use a pulley where the antenna wire is connected to one of the trees. A nylon rope threaded through the pulley drops down to a 50 pound weight (concrete bricks) which is suspended in midair. When the wind bends the trees apart, the weight will rise and the antenna wire will not snap.

TRANSMATCH

We have seen how important it is to match antennas to the impedance of their feedline. A given antenna length is most desirable at different frequencies. If it is impractical to cut the antenna to the ideal length, then we can compensate by using a device called a transmatch. This is a small box that can tune any random length of wire to the proper electrical length for use on any amateur band. The transmatch consists of a large variable inductor and several large capacitors which are balanced to ground. Suppose we have a 42 foot antenna and we wish to use it on 80 meters where quarter-wavelength is some 60 feet. The transmatch variable inductor will act as part of our 42 foot wire, the additional inductance making 42 feet look like 60 feet of wire to the RF energy. The earth will serve as the other quarter-wavelength, and the 42 foot wire connected to the transmatch will accept RF from the transmitter. Suppose, on the other hand, we have a 36 foot wire which we hope to use on 40 meters. A quarter-wavelength on 40 meters is only 30 feet. The balanced capacitors in the transmatch are placed in series with the 36 foot wire. They are tuned so that they effectively shorten the 36 foot wire. When the 36 foot wire looks like 30 feet at 40 meters, it will accept RF from the transmitter. In addition to accepting RF, the tuned wire will reject harmonics or multiples of the transmitter's frequency. Thus the transmatch is also an aid in eliminating undesirable harmonics. If you were transmitting on 3.7 MHz, your second harmonic would appear on 7.4 MHz and interfere with stations in that part of the spectrum, which is not an amateur frequency.

6

A Practical Guide to Schematic Diagrams and Basic Electron Theory

Many electronic components resemble each other in physical appearance. If we draw a blueprint of an electronic circuit using a picture of the component, we would soon go astray. This is why the schematic diagram has been devised. There can be no mistaking one component for another. Figure 6-1 illustrates the most often used electronic symbols.

Lines which represent wiring or connections are drawn from one component to another. When two intersecting lines are joined with a dot, these two wires connect to each other. However, when one intersecting line loops around the other, the two wires are not connected.

As Figure 6-1 shows, certain capacitors show a polarity sign (+ or −). If the circuit's wiring is not connected to the correct polarity, the circuit will not work. All diodes have a cathode and an anode. A colored band will be painted on the cathode side of the diode. When using a diode as a detector, the cathode side is at the output. When using transistors, the PNP and the NPN transistor cannot be interchanged. The PNP collector takes a negative voltage, while the NPN collector takes a positive voltage.

A coil will usually have two leads and it is not necessary to identify each. If there is a third lead, it will be a tap and it should

SCHEMATIC SYMBOLS USED IN CIRCUIT DIAGRAMS

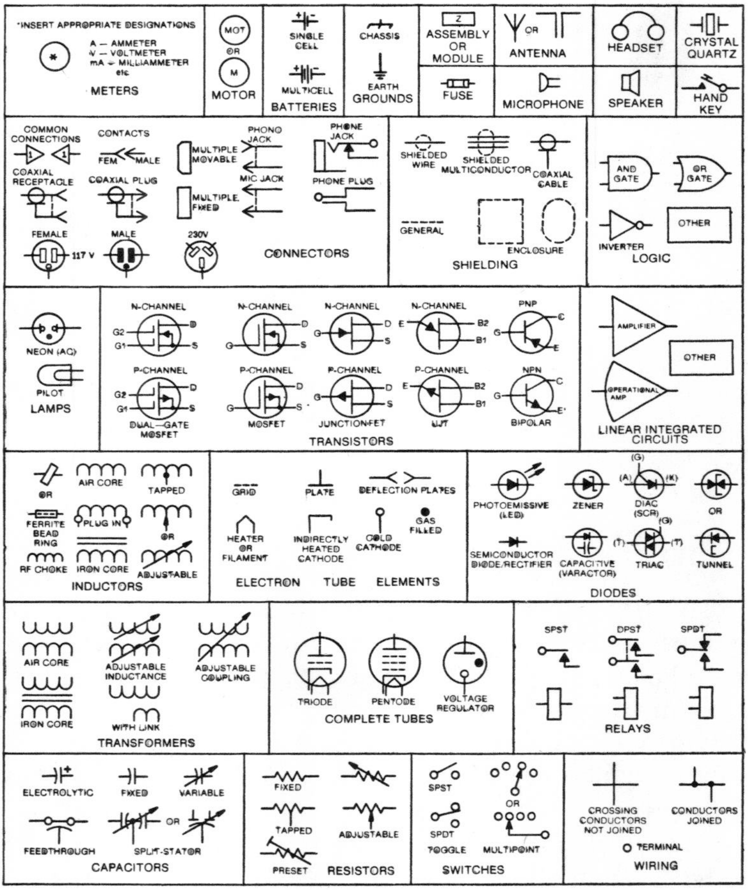

Figure 6-1 *Symbols used in circuit diagrams.*

be identified. A transformer will have many leads. They should be color coded for identification, as shown in Figure 6-2. To verify the color code, it is always a good idea to measure the continuity across two leads with an ohmmeter. A very low resistance (0-5 ohms) between two leads will indicate they both are connected to the same inductor or coil of wire within the transformer. A coil usually offers inductance to tune a resonant circuit, or it might be a choke to stop high frequencies from passing through it.

SCHEMATIC DIAGRAMS AND BASIC ELECTRON THEORY 115

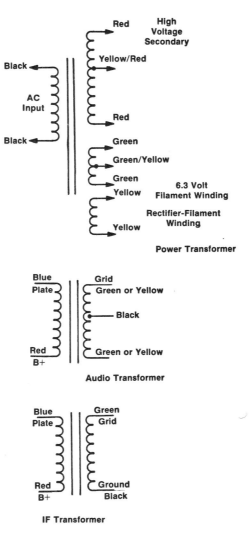

Figure 6-2 *Color code identification for power transformers, audio transformers, and IF transformers.*

ELECTRON THEORY

One third of the questions on the Novice FCC examination will be on basic electron theory. You will have to know what voltage means, what current means, and what resistance means. You will have to know how to solve simple Ohms Law

116   SCHEMATIC DIAGRAMS AND BASIC ELECTRON THEORY

problems, the principles of electromagnetism, reactance, and impedance. In addition, you will be required to know the principles of a rectifier, an amplifier, and an oscillator.

The illustrations and captions which follow should give the

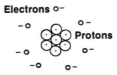

Atoms contain electrons and protons. Electrons carry a negative charge and can move from atom to atom. Protons carry a positive charge and normally do not move from atom to atom.

Like charges repel, opposite charges attract.

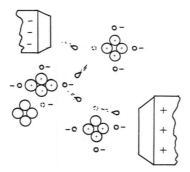

When a strong positive and negative charge are applied to atoms with loosely held electrons, the electrons will move from atom to atom. This is the basic principle of current flow.

Figure 6-3

SCHEMATIC DIAGRAMS AND BASIC ELECTRON THEORY   117

reader an excellent understanding of the topics mentioned above. This information has been reprinted from the Official Novice Student Workbook, published as part of the ARRL Training Program.

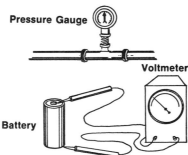

Water pressure is measured by a pressure gauge. Electrical pressure is also measured with a pressure gauge. Since electrical pressure is called voltage, the electrical pressure gauge is called a voltmeter. Electrical pressure is measured in units called volts.

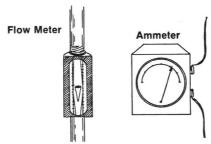

The amount of water flowing through a given point during a given time period is called the flow rate, and is measured in units called gallons per minute. Electrical current is also measured in terms of flow rate, and is measured in units called amperes. Water flow rate is measured with a flow meter. Electrical current flow is also measured with a flow meter, called an ammeter.

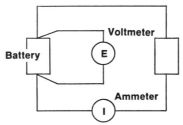

Voltage is measured across the voltage source. Current is measured in the circuit so that it must pass through the ammeter.

Figure 6-4

SCHEMATIC DIAGRAMS AND BASIC ELECTRON THEORY

PROPERTY	ABBREV.	SYMBOL	UNIT OF MEASURE		
RESISTANCE $R = \frac{E}{I}$	R	—⋀⋀⋀— R	·OHM ·KILOHM ·MEGOHM	$R \times 1$ $R \times 1,000$ $R \times 1,000,000$	Ω kΩ MΩ
CAPACITANCE	C	—)|— C	·FARAD ·MICROFARAD ·PICOFARAD	$C \times 1$ $C \times .000,001$ $C \times .000\,000,000,001$	F μF pF
INDUCTANCE	L	⌒⌒⌒ L	·HENRY ·MILLIHENRY ·MICROHENRY	$L \times 1$ $L \times .001$ $L \times .000,001$	H mH μH
VOLTAGE $E = IR$	V	E	·VOLT ·KILOVOLT ·MILLIVOLT ·MICROVOLT [VOLTMETER]	$V \times 1$ $V \times 1,000$ $V \times .001$ $V \times .000,001$	V kV mV μV
CURRENT $I = \frac{E}{R}$	A AMP	I	·AMPERE ·MILLIAMPERE MICROAMPERE [AMMETER]	$A \times 1$ $A \times .001$ $A \times .000,001$	A mA μA
POWER $P = EI$	P	P	·WATT ·KILOWATT [WATTMETER]	$W \times 1$ $W \times 1000$	W kW

Figure 6-5

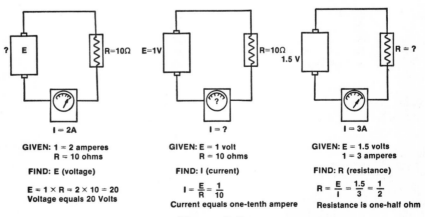

Figure 6-6

SCHEMATIC DIAGRAMS AND BASIC ELECTRON THEORY

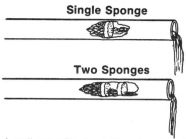

A small sponge filter inserted in a pipe reduces the amount of water flowing by 50%. If another identical sponge were added, the water flow rate would be reduced by another 50%. Thus, adding more sponges through which the water must pass reduces the amount of water flowing.

Resistance limits the amount of current that can flow. Adding a second resistance further reduces the current flow just as adding a second sponge reduced the water flow. Thus, adding resistances in series reduces current below that which would flow with only a single resistor of equal value.

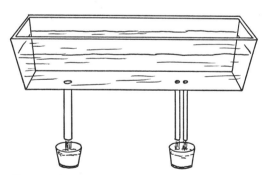

Two water pipes can together carry more water per minute than can a single pipe of the same diameter. That means that more water will flow through two parallel pipes than will flow through a single pipe of the same diameter.

Figure 6-7

SCHEMATIC DIAGRAMS AND BASIC ELECTRON THEORY

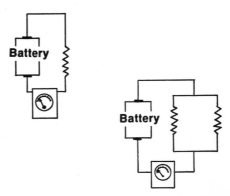

The current flowing through each of the three resistors above would be determined by Ohm's Law. If the resistances were the same, and the battery voltage was the same, the current flowing through each resistor would be the same as that flowing through the others. The upper battery would provide current for only one resistor while the lower battery would be providing current for two resistors. That means twice as much current would flow in the lower circuit. Thus, more current flows through two parallel resistors than flows through a single resistor of the same value.

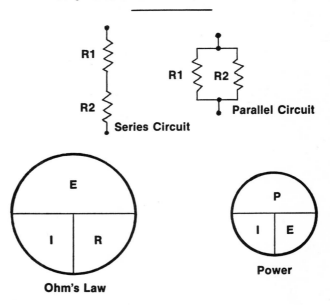

Figure 6-8

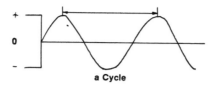

Figure 6-9

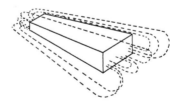

Figure 6-10

A permanent magnet has a magnetic field around it that can be thought of as a large number of magnetic lines of force that extend outward from the poles.

An electromagnet is produced by passing current through a wire wound around an iron core. The magnetism produced remains only as long as the current is made to flow through the wire.

Lines of Magnetic Force

Even if the iron core is removed, a magnetic field just like that around a permanent magnet is generated around a coil when current passes through it.

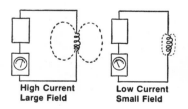

High Current
Large Field

Low Current
Small Field

The strength of the magnetic field around a coil depends on the amount of current flowing through the coil. The greater the current, the greater the magnetic field.

Figure 6-10 con't.

SCHEMATIC DIAGRAMS AND BASIC ELECTRON THEORY

AVERAGE CURRENT

When voltage is applied across a coil, current does not instantly reach its maximum value. Instead it slowly builds up as shown by the left portion of the solid curve. If the voltage remained unchanged, the current would continue as shown by the dotted line. However, if the voltage were removed, the current would decay in the same manner as it built up, as shown by the center portion of the solid line. If the voltage were alternately applied and removed, or an AC voltage were applied, the wave shape would look like the entire solid line. If you averaged the current waveshape, you would have an average current as shown by the horizontal dashed line.

LOW FREQUENCY AVERAGE
HIGH FREQUENCY AVERAGE

If the frequency of the AC voltage were increased, the time between each alternation would be much less than it would be at a much lower frequency. This means the current would build up to a smaller value before beginning to decay. If you compared the average current flowing when a high frequency is applied to the average current flowing when a low frequency is applied, it would be much lower.

INDUCTIVE REACTANCE

CURRENT

A graph of the average current flowing through the coil plotted against the frequency of the applied voltage would look like the solid line. The dashed line represents the inductive reactance, which behaves like "resistance" in limiting current. As the frequency increases, the inductive reactance increases causing the average current to decrease.

Figure 6-11

QUIZ

1. What is the major difference between AC and DC?

2. What is a cycle?

3. What is frequency?

4. What is the abbreviation for frequency?

5. What is frequency usually measured in? Abbreviated?

6. What is the approximate range of audio frequencies? Radio frequencies?

7. What is the approximate frequency of the am broadcast band?

8. Can DC be affected by inductance? Explain.

9. Why does the inductive reactance increase with frequency?

10. How can the inductive reactance be increased besides by increasing frequency?

11. What can a transformer do?

12. What is the purpose of a choke?

Figure 6-12

SCHEMATIC DIAGRAMS AND BASIC ELECTRON THEORY

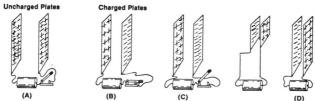

If two metal plates are placed facing each other and close together, but not touching, as in (A), are connected to a battery, as shown in (B), one plate will lose all of its free electrons and become positively charged. The other plate will accumulate an excess of electrons and become negatively charged. Once charged, the plates will retain their charge until discharged, even though the battery is disconnected, as shown in (C). The amount of charge that can be stored depends on how close together the plates are and on their size. All things being equal, the larger the plates, the greater the charge, as shown in (D).

Figure 6-13

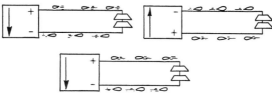

If the polarity of the applied voltage is reversed, the electrons on the negatively charged plate will flow through the circuit to the other plate, making it become negatively charged. If an AC voltage is applied, the electrons will flow through the circuit from one plate to the other, back and forth as the polarity of the voltage alternates. The effect on the other parts of the circuit is the same as if current actually flowed through the capacitor, which it does not.

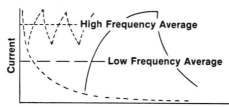

When a voltage is applied to a capacitor, a large number of electrons begin to flow into the negative plate. As time goes by and the plate becomes more negatively charged, the number of electrons flowing into it decreases. If the number of electrons flowing was plotted against time, it would look like the left portion of the solid line. If the voltage remained unchanged, the current curve would continue as shown by the dashed line. If a low frequency AC voltage is applied, the curve would look like the entire solid line. The average current would be as shown by the horizontal dashed line. If the frequency is increased, as shown by the dotted lines, the average current would be higher.

Figure 6-14

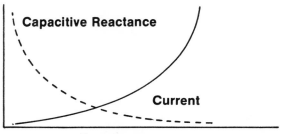

If you plot the average current flowing through the capacitor circuit against the frequency of the applied voltage, the curve would look like the solid line. The current limiting factor, called capacitive reactance, would appear as the dashed line on the graph.

Figure 6-14 con't.

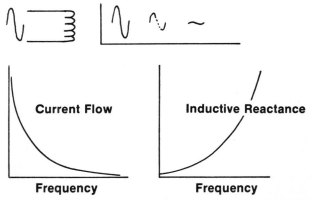

If an AC voltage is applied to an inductor, the current flowing through the circuit will decrease as the frequency is increased.

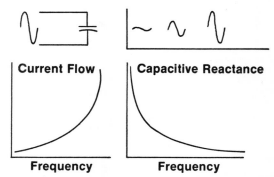

If an AC voltage is applied to a capacitor, the current flowing through the circuit will increase as the frequency increases.

Figure 6-15

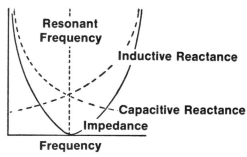

Resonance is the frequency at which the capacitive reactance and the inductive reactance are equal. Since they cancel out, the net impedance is zero.

Figure 6-15 con't.

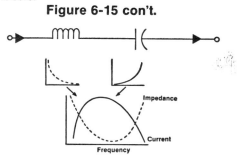

At low frequencies the capacitive reactance is high and the current flowing through it is low. At low frequencies the inductive reactance is low and current flowing through it is high. At high frequencies it is just the opposite. Maximum current flows when the reactances are equal. As the frequency increases or decreases, the inductive or capacitive reactance increases, reducing the current flowing through the series circuit.

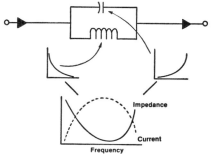

At low frequencies, the capacitive reactance is high and the current flowing through it is low. At low frequencies the inductive reactance is low and current flowing through it is high. At high frequencies, it is just the opposite. Maximum current flows at very low frequencies (through the inductor) and at very high frequencies (through the capacitor). Minimum current flows at resonance when the current flow through each is equal.

Figure 6-16

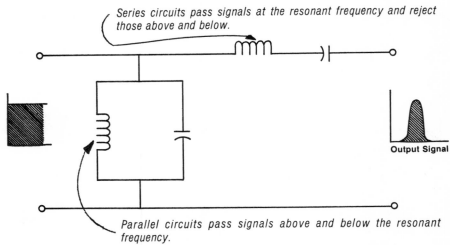

Series circuits pass signals at the resonant frequency and reject those above and below.

Output Signal

Parallel circuits pass signals above and below the resonant frequency.

Series and parallel tuned circuits can be used to reduce the level of undesired frequencies in a signal.

Figure 6-16 con't.

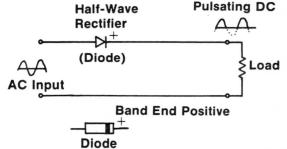

Half-Wave Rectifier

Pulsating DC

(Diode)

Load

AC Input

Band End Positive

Diode

The basic rectifier circuit consists of a single diode that operates as a one-way valve. Electrons can pass through it only from the cathode (banded end) to the anode. When the anode is positive in respect to the cathode, the diode acts like a conductor. When the cathode is positive in respect to the anode, it acts like an open circuit.

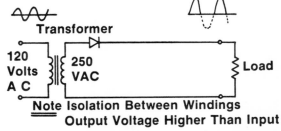

Transformer

120 Volts AC

250 VAC

Load

Note Isolation Between Windings
Output Voltage Higher Than Input

The same circuit can be isolated from the voltage source, and the output voltage changed by inserting a transformer between the source and the rectifier circuit.

Figure 6-17

SCHEMATIC DIAGRAMS AND BASIC ELECTRON THEORY

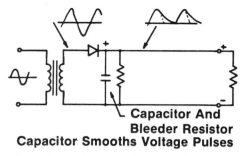

Capacitor And Bleeder Resistor
Capacitor Smooths Voltage Pulses

The pulsating DC can be somewhat smoothed by connecting a capacitor across the circuit as shown. To prevent the possibility of being shocked when the power is disconnected, a bleeder resistor is connected across the capacitor. This resistor safely discharges the capacitor—it bleeds off the charge.

Figure 6-17 con't.

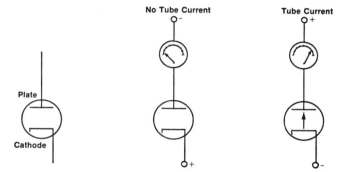

When the plate is negative in respect to the cathode, no current flows through the tube. However, if the plate is positive in respect to the cathode, current does flow through the tube, from cathode to plate.

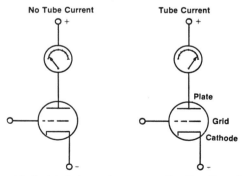

When the grid of a triode is much more negative than the cathode, it prevents electrons from flowing to the plate and no current flows through the tube. When the grid is at the same voltage as the cathode the tube behaves like a diode and current flows through the tube.

Figure 6-18

SCHEMATIC DIAGRAMS AND BASIC ELECTRON THEORY

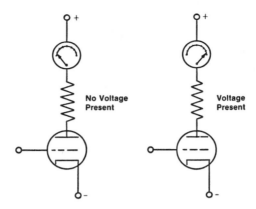

When a resistor is connected between the plate and the positive voltage, all tube current must flow through it—it is in series with the tube. When the grid prevents current from flowing, there is no voltage developed across the resistor. When the grid lets current flow, the current flowing through the tube produces a voltage across the resistor in accordance with Ohm's Law.

Figure 6-18 con't.

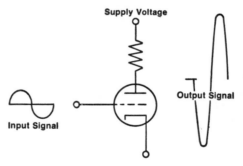

When the voltage on the grid changes because of a signal, the current flowing through the tube changes as well. The changing current flowing through the tube produces a changing voltage across the resistor. If you measure the voltage on the plate of the tube, it will be equal to the supply voltage minus the voltage developed across the resistor. Thus, when the input signal is going more negative and causing the tube current to decrease, the plate voltage is increasing (supply voltage minus a smaller resistor voltage). Because of this, the output is inverted from the input.

Figure 6-19

SCHEMATIC DIAGRAMS AND BASIC ELECTRON THEORY   131

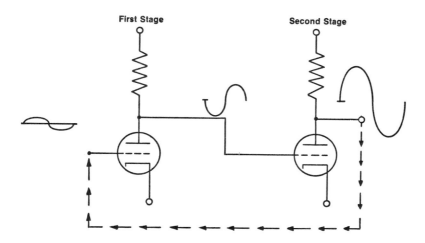

The output of a "two stage" amplifier is in-phase with the input signal. If the gain is high enough, some of the output can feed back into the input and be reamplified—causing oscillation.

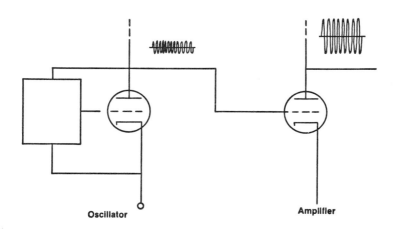

If the output of an oscillator is connected to the input of an amplifier, the output of the amplifier will be an enlarged replica of the oscillator output. This is the basis of all modern transmitters.

Figure 6-19 con't.

SCHEMATIC DIAGRAMS AND BASIC ELECTRON THEORY

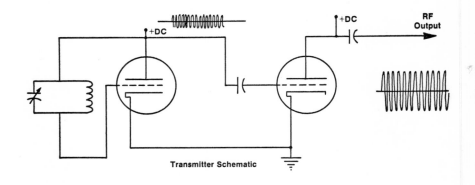

Transmitter Schematic

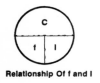

Relationship Of f and I

Relationship of I and f			
I	Approximate f	I	Approximate f
160 meters	2 MHZ	15 Meters	20 MHZ
80 meters	4 MHZ	10 Meters	30 MHZ
40 meters	7 MHZ	6 Meters	50 MHZ
20 meters	15 MHZ	2 Meters	150 MHZ

Figure 6-20

7

Working with Semiconductors and Other Components

SEMICONDUCTOR THEORY

A conductor favors the flow of electrons, while an insulator impedes the flow of electrons. Halfway between the two are semiconductors. These may be silicon or germanium. Their advantage is small size, long life, and low power requirements. Conductors have many free electrons while semiconductors have few free electrons. When carefully controlled amounts of impurities (materials with a different atomic structure) such as arsenic or antimony are added to a semiconductor, more free electrons are created. However, when other impurities are added, such as indium, aluminum, or gallium, a deficiency or hole is produced. This increases conductivity. Semiconductors that conduct by the action of free electrons are called N-type material; material that conducts by the action of an electron deficiency is called P-type material.

JUNCTIONS

When P-type and N-type materials are joined, a rectifier is created. Most of the current flows in the forward direction; however, a small amount of reverse current flows in the undesired direction. A semiconductor's junction has almost no spacing, therefore, a large capacitive effect (capacitor) is created across the semiconductor. This effect limits the operating frequency of a semiconductor as compared to vacuum tubes. How the excess electrons and holes recombine in the material depends on temperature. Thus semiconductors are extremely temperature sensitive. It is possible to reduce capacitance by making the contact area very small. A point contact forms the P-type region when N-type material is used for the main body of the device.

DIODES

Back in 1904, the galena diode was patented. This device detected radio waves, converting them into audio frequencies. Remember how this crystal diode became the heart of the crystal receiver, and the crystal receiver in turn became the grandfather of modern day receivers. After the vacuum tube was invented, the galena (crystal) detector became obsolete. The vacuum tube detector could be inserted inside a vacuum tube serving yet another function. However, when transistors came into widespread use, the vacuum tube detector itself soon became obsolete, for it required filament power, was bulky, and was subject to filament burn-out. The solid-state diode may be a point-contact or junction-type. Figure 7-1 illustrates the diode symbol, the point-contact diode, and the junction-type diode. Silicon is used for diodes which operate at very high frequencies and in the microwave regions where heat can be a problem. Germanium diodes are the most sensitive. They also have a small current flow in the reverse direction.

When it is necessary to rectify high currents up to 50 amperes, the silicon junction-type diodes are used. They can withstand reverse peak voltages up to 2500 volts and can be connected in parallel to increase their current handling capability. Silicon diodes do not make good detectors of very small signals, for they require a forward voltage of 0.4 to 0.7 volts.

Semiconductor diodes are rated in PIV (max safe inverse voltage or peak inverse voltage), and PRV (max average rectified current). The Zener diode is a device that has a sharp break from nonconductance to conductance. The voltage point at which this break takes place is called the Zener knee. As can be seen in Figure 7-2, when the forward voltage begins to

SEMICONDUCTORS AND OTHER COMPONENTS

exceed 0.75 volts, the I_f (forward current through the Zener diode) starts to rise rapidly. When a Zener diode is placed across a voltage, it will conduct when its Zener knee (Zener voltage) is exceeded. Thus it becomes a very good voltage regulator and transient protector. A transient is an unwanted voltage spike that can damage transistors.

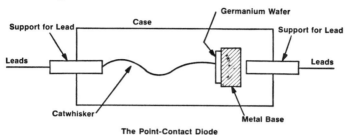

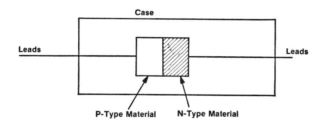

Figure 7-1 *The germanium point-contact diode, and the silicon junction diode.*

"Knee" curve of a 6 volt Zener diode

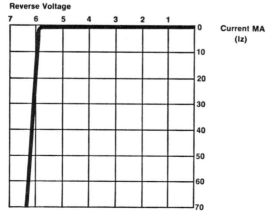

Figure 7-2 *Zener diode characteristics. The voltage drop is constant at 6 volts in the reverse direction. A sudden current increase then occurs.*

TRANSISTORS

When two layers of P-type semiconductor material surround a thin layer of N-type material, a PNP junction has been formed. When two layers of N-type material surround a thin layer of P-type material, an NPN junction has been formed. Figure 7-3 shows the PNP transistor with its PNP junction. As can be seen, the thin layer forms the base, the two surrounding layers form the emitter and the collector. The important thing to remember about transistors is that the collector voltage on the PNP transistor must be negative, while the collector voltage on the NPN transistor must be positive. In Figure 7-4 we see that the emitter and collector form the transistor's output circuit, while the emitter and the base form the input circuit.

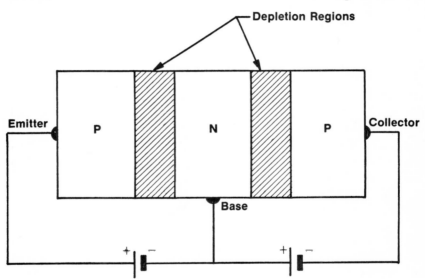

Figure 7-3 *The junctions inside the PNP transistor. Ion interchange occurs within the depletion regions. The battery polarity must be observed.*

The collector-emitter circuit is biased in the reverse direction so only a small current can flow. The emitter-base junction is biased in the forward direction by the input signal. The small input signal (input circuit) can control the larger current in the emitter-collector output circuit. Thus the transistor is capable of amplification. If some of its output

SEMICONDUCTORS AND OTHER COMPONENTS   137

The transistor as an amplifier or a current control valve.

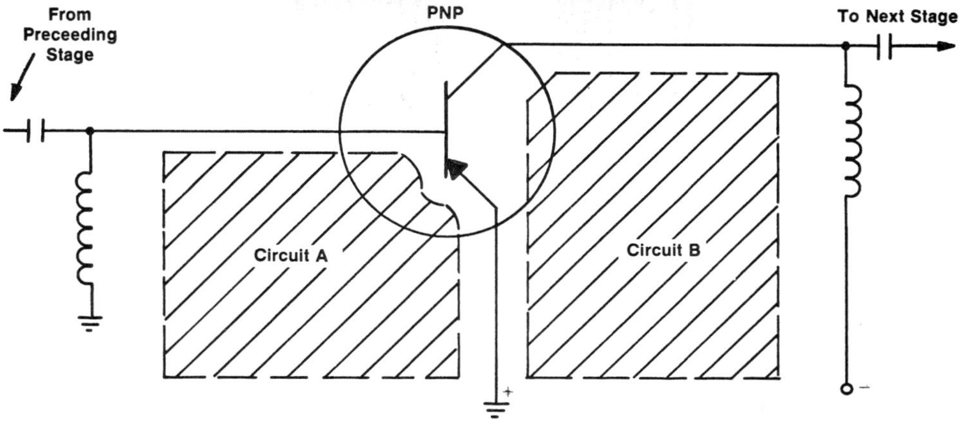

Figure 7-4 *Think of a transistor circuit as 2 separate circuits; an input circuit A, and an output circuit B. The current in A is delivered from the preceeding stage; the current in B is many times greater than A. This explains the amplification that occurs.*

circuit energy is fed back to the input circuit, then it becomes an oscillator. This circuit is seen in Figure 7-5. The quartz crystal feeds energy from the collector circuit back to the input (base) circuit. This energy is in-phase, or adds to the input energy so that the entire circuit oscillates or generates a sine wave. The frequency of the sine wave is determined by the quartz crystal.

QUARTZ CRYSTALS

Several crystalline substances found in nature can convert mechnical strain into an electrical charge and vice versa. This is called the *piezoelectric effect.* When a voltage is placed across a thin plate of quartz, the piezoelectric effect produces mechanical movement which in turn creates another voltage. This crystal output voltage is a sine wave, the frequency of which depends on how thick the quartz plate is, its shape, and its dimensions. The phonograph crystal converts mechanical strain (from the record groove) into the input signal to your hi-fi set. The quartz crystal is actually a tuned circuit (LC) having both inductive and capacitive reactance. The frequency where the reactances peak becomes the crystal's resonant frequency.

SEMICONDUCTORS AND OTHER COMPONENTS

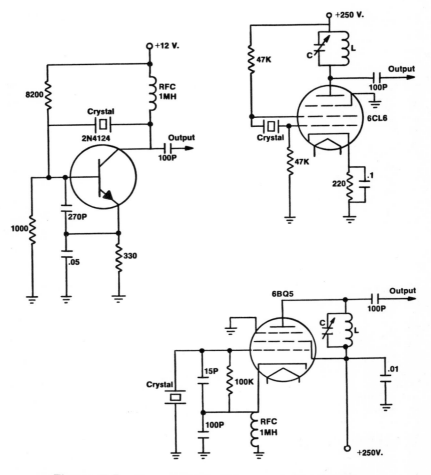

Figure 7-5 *These crystal oscillators are designed to deliver power to high powered amplifiers. The LC circuits can resonate at harmonics of the crystal frequency.*

FIELD EFFECT TRANSISTORS

The resistance as impedance across the input of a transistor is extremely low. When high impedance devices are used this becomes a problem. It is impossible to match high impedance circuits into low impedance circuits. With the invention of the FET and the MOSFET, this problem was solved. Both FET and MOSFET devices have an insulated gate instead of a base as the input circuit. The insulated gate is separated from the rest of

SEMICONDUCTORS AND OTHER COMPONENTS

the device by a thin dielectric or insulator. This accounts for the high input impedance of these devices, which may go up to a million megohms. A dual gate MOSFET has two high impedance gate inputs. One can be used as a signal input and the other connected to an AGC signal. AGC (automatic gain control) will decrease the sensitivity of an amplifier when a very strong signal is received so that no distortion can occur.

RESISTORS

Carbon resistors are satisfactory for most applications. They come in 5 and 10 percent tolerances. When high precision resistors are required for critical applications, then wire-wound resistors may be used. Tolerances of 1 percent can be achieved this way. A critical parameter in amateur radio circuits is the wattage of a resistor. The required wattage of a resistor is not difficult to determine.

Remember Ohm's Law. A derivation of the Ohm's Law formula is $P = I^2R$, or power (in watts) is obtained by multiplying the square of the current by the value of the resistance in ohms. Approximate values of current are usually known without even using a milliameter to measure the current. If a resistor is placed in the plate circuit of a vacuum tube, we can look at a radio or TV tube manual to find the current flowing in the tube's plate circuit. Let's suppose we find that 20 milliamperes is flowing in the plate circuit—or .02 amperes. If the resistor has a value of 1000 ohms, then $(.02)^2 \times 1000 = .0004 \times 1000 = 0.4$ watts. Resistors are sold in ratings of one-quarter watt, one-half watt, one watt, two watts, five watts, ten watts, and higher. So our calculation told us that 0.4 watt would be going through our resistor. Then we would use a half-watt resistor in our circuit.

CAPACITORS

Figure 7-6 shows typical capacitors. Capacitors can be used to filter the pulsating DC that comes out of a rectifier. These are electrolytic capacitors. Their correct polarity must be observed. Their voltage rating should be 25 percent above the voltage in that circuit.

Capacitors used to bypass RF in receiver and transmitter circuits are usually ceramic capacitors. Their voltage rating should always exceed the voltage in the circuit. They are sometimes called disk capacitors. Their value is usually from 0.1 mfd to .001 mfd. Disk ceramic capacitors are also used to couple RF and audio signals. Values here may be from 100 mmfd to .01 mfd.

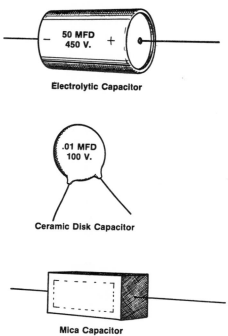

Figure 7-6 *Typical capacitors: the electrolytic which is polarized, the ceramic and the mica which are not polarized.*

Disk ceramic capacitors have a higher loss at high frequencies than do mica capacitors. Thus in oscillator circuits and RF receiving circuits, mica capacitors are recommended. Their values range from 10 mmfd to 1000 mmfd, which is the equivalent of .001 mfd. The abbreviation SM for silver mica denotes a capacitor of higher quality (Q) at the higher frequencies than even mica. Silver micas are often used to provide a feedback path in crystal oscillators. You will see an example of this in Chapter 11; there, a homemade transmitter is described. Always remember that a capacitor will pass AC and the high frequencies, but will block DC.

CHOKES

Figure 7-7 shows typical chokes. Chokes are devices that pass DC but block audio or RF frequencies. An audio choke (stops audio) has an iron core and resembles a small power transformer or audio transformer. On the other hand, RF chokes are small. They may consist of a single winding of very fine wire on a quarter-inch insulated rod. These chokes would have inductance values of 10 microhenrys up to 100 microhenrys. Larger RF chokes can handle several hundred milliamperes and have inductances up to 5,000 microhenrys or 5 millihenrys. These chokes usually have two to five separate pi-windings. This minimizes capacitance leakage, which would negate the purpose of the choke. In the transmitter circuit of

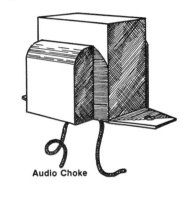
Audio Choke

RF Choke (Low Inductance)

RF Choke (Large Inductance)

Figure 7-7 *Typical chokes.*

Figure 11-2 (see p. 198), there are three RF chokes. RFC1 passes DC to the oscillator's cathode but blocks the feedback energy from being grounded. RFC2 passes the DC plate voltage but blocks the oscillator RF output. RFC3 passes DC to the power amplifier plate but prevents the RF from being grounded.

RF INDUCTORS

As we have seen, RF chokes have a very high reactance (opposition) to the flow of high frequency energy. They are broadband devices. In other words, they will block RF energy over a broad part of the spectrum. An RF inductor, as distinguished from an RF choke, is not broadband. It has a given inductance. Every piece of wire has inductance when AC flows through it, as we discovered in the previous chapter. As we increase the number of turns of a coil or increase the diameter of the coil, the inductance increases. If we connect a capacitor across the ends of the inductor or coil, then we have a tuned circuit.

Remember the LC circuit of the crystal radio in Chapter 4. The LC (inductor and capacitor) combination did not pass RF energy at a given frequency. It is tuned to or resonant to that frequency. The resonant frequency is where the capacitive and inductive reactance is equal. As they oppose each other equally, no current is able to flow through the circuit. Table 7-1 shows how to find the unknown value of the RF inductor (in microhenrys) or how to find the unknown value of the capacitor (microfarads), when we wish an LC combination to tune to a known frequency in kilocycles or megacycles. For example, suppose we wish an LC combination to tune to the 40 meter amateur band (7 MHz), and we have a 10 microhenry (10 Mh) inductor in the circuit. We wish to find out what value of capacitor will be required to tune this LC combination to 7 MHz.

Looking at the frequency divisions at the bottom of the table, a 10 Mh heavy line rises upward to the right, starting near 16 kilocycles. Follow this (X_L) inductive reactance line up until it intersects with the vertical line which represents 7 megacycles. Making a mark at the intersection, determine the nearest capacitive reactance line which moves downward to the right.

SEMICONDUCTORS AND OTHER COMPONENTS

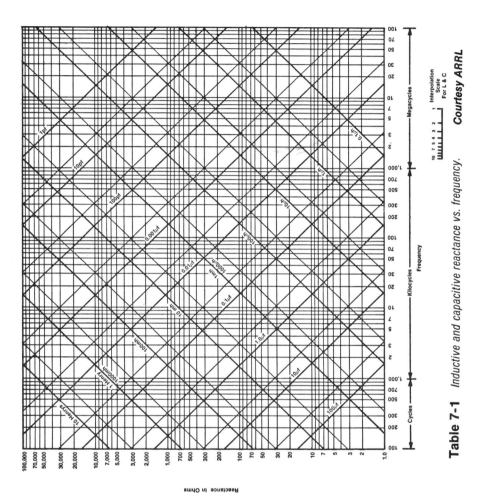

Table 7-1 *Inductive and capacitive reactance vs. frequency.* *Courtesy ARRL*

Inductive and capacitive reactance vs. frequency. Heavy lines represent multiples of 10, intermediate light lines multiples of 5; e.g., the light line between 10 µH and 100 µH represents 50 µH, the light line between 0.1 µF and 1 µF represents 0.5 µF, etc. Intermediate values can be estimated with the help of the interpolation scale.

Reactances outside the range of the chart may be found by applying appropriate factors to values within the chart range. For example, the reactance of 10 henrys at 60 cycles can be found by taking the reactance to 10 henrys at 600 cycles and dividing by 10 for the 10-times decrease in frequency.

Courtesy ARRL

The point we have marked lies on the 50 PF (micro-microfarad), X_c line. This line is between the 10 PF and the 100 PF lines.

We have discovered that a 10 microhenry inductor across a 50 PF micro-microfarad capacitor will tune to 7 MHz or 40 meters. We may wish to compensate for external inductance (wiring) and capacitance, also to alter the resonant frequency. To do this, we would make the capacitor variable (a tuning capacitor). In this example, we have discovered that a variable capacitor of 2-100 mmfd will certainly allow our circuit to tune to 7 MHz. Another interesting illustration of this concept is the pi-tuning network in Figure 11-2 consisting of L1, C4, and C5. This circuit is seen in Figure 7-8 (see below). When these components are tuned to resonance, the maximum power from the transmitter will be radiated on this frequency. Remember, however, that in this circuit C4 and C5 are in series. When two capacitors are in series, their total capacitance is less. If 50 micro-microfarads of capacity is required, then C4 and C5 must each be 100 mmfd, or they may be any other combination that will produce a total capacity of 50 micro-microfarads. To calculate the total capacity of two series capacitors use this formula:

$$\frac{C1 \times C2}{C1 + C2} = \text{total capacity}$$

Where a 100 mmfd and a 50 mmfd capacitor are in series, then:

$$\text{total C} = \frac{100 \times 50}{100 + 50} = \frac{5000}{150} = 33 \text{ mmfd}$$

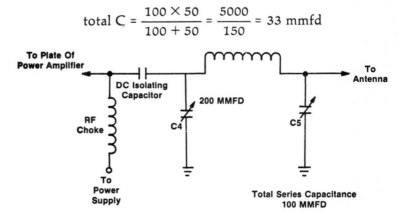

Figure 7-8 *The pi-tuning network. Part of the tank circuit of the transmitter of Figure 11-2. As C4 and C5 are both grounded, they are across L1 in a series arrangement.*

SEMICONDUCTORS AND OTHER COMPONENTS

When capacitors are in parallel (across each other), simply add their individual capacity together to obtain the total capacity of the circuit.

8

Simplified Guide to Amateur Circuits in Detail

HIGH-PASS, LOW-PASS, AND BAND-PASS FILTERS

In Chapter 6 we found that the reactance of both capacitors and inductors varies with frequency. As the frequency increases, the reactance of an inductor increases. Also, as the frequency increases, the reactance of a capacitor decreases. Thus if we wish to favor the higher frequencies, we would insert capacitors in series with the flow of current. On the other hand, if we wished to favor the low frequencies, we would place capacitors across the current path to ground. Conversely, if we wished to favor the high frequencies, we would place an inductor across the current path to ground (the high frequencies would not get through). On the other hand, if we wish to favor the low frequencies, we would place an inductor in series with the current path. The most complex filters are based on these simple principles of how capacitors and inductors discriminate against particular frequencies. In Figure 8-1 we see a simple illustration of this principle. This is called a one section high-pass filter.

HIGH-PASS FILTERS

Capacitor C1 is in series with the flow of current. It favors the high frequencies, while discriminating against the low frequencies. Inductors L1 and L2 easily short circuit low frequencies to ground, while having a high reactance to high frequencies, thus allowing them to pass.

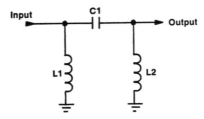

Figure 8-1 *The high-pass filter.*

LOW-PASS FILTERS

The converse of this circuit is seen in Figure 8-2. This is a low-pass filter. The high frequencies are blocked by the reactance of L1. The high frequencies are also short-circuited to ground by C1 and C2. Only the low frequencies are able to pass through this circuit.

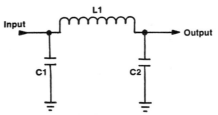

Figure 8-2 *The low-pass filter.*

To prevent amateur interference to a TV set (TVI), you might insert a high-pass filter at the antenna terminals of a TV set. Thus the low frequency amateur frequencies would be blocked from the TV set, while the high frequency TV stations would still get through. Another preventive measure would be to install a low-pass filter at the antenna terminal of your amateur transmitter. The low frequency amateur frequencies would pass through this filter easily, going to the antenna. However, harmonics and other higher frequencies would be blocked and would never get to the antenna.

The same filter principles can be used at audio frequencies. In this case we call them emphasis or de-emphasis filters. For

AMATEUR CIRCUITS IN DETAIL

example, suppose you were told that your voice (over ham radio) had a very unpleasant bass quality without the highs coming through. You would have to add a de-emphasis circuit in the audio section of your transmitter. This high-pass filter would block the low frequencies (the bass voice frequencies) allowing the higher voice frequencies to pass. This modification would give your voice a much more pleasant quality and also improve your communication ability, as the higher voice frequencies, which are crisper, are more easily heard under noisy conditions.

BAND-PASS FILTER

It is also possible to put a high-pass filter together with a low-pass filter to form a band-pass filter. Let's suppose that we wished to pass only the 80 meter amateur band from 3.5 to 4 MHz. We could thus improve the quality of reception on this band, because annoying interference existed below 3.5 MHz. and above 4 MHz. We would then build a high-pass filter with its cut off frequency at 3.5 MHz. This filter would suppress all frequencies below 3.5 MHz. Next, we would add a low-pass filter with its cut off frequency at 4 MHz. This filter would suppress all frequencies above 4 MHz. Thus frequencies below 3.5 MHz are cut off, while frequencies above 4 MHz are also cut off. We have built a band-pass filter which passes only a given band of frequencies—from 3.5 MHz to 4 MHz.

The band-pass filter is also used in SSB. A common example of this is the sideband filter in an SSB transmitter. Here, a typical sideband filter would pass 300 Hz to 3000 Hz away from the carrier frequency, which is the most vital portion of voice frequencies as far as communication quality is concerned. This single sideband passes through the filter. The carrier frequency, which corresponds to zero hertz, does not pass. Also, the remaining sideband is outside the pass-band of this band-pass filter. It is then suppressed so that only a single sideband passes through to the RF power amplifier.

Figure 8-3 shows a crystal filter used to pass a single sideband. It consists of one balanced section composed of the crystals Y1, Y2, Y4, and Y5. Also, the second balanced section consists of the crystals Y3 and Y6. The crystals Y1, Y2, and Y3

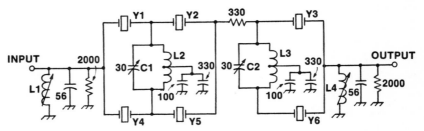

Figure 8-3 *Diagram of a two-section crystal filter used in a single sideband.*

resonate at the carrier frequency, offering the highest impedance at this frequency. The carrier frequency is thus suppressed. The crystals Y4, Y5, and Y6 are resonant 1500 Hz away from the carrier frequency. They suppress the unwanted sideband. In this particular type of band-pass filter, all the unwanted frequencies are rejected. Only the desired hand (a single sideband) can pass. This type of filter is also called a mechanical filter because its major components (quartz crystals) vibrate mechanically to establish their resonant frequency.

POWER SUPPLIES

With the exception of battery operated equipment, all amateur gear requires some type of power supply. It is this circuit that changes AC line power into voltages that can be used for vacuum tubes, transistors, and other electronic components.

The simplest power supply circuit pictured in Figure 8-4 is the half-wave rectifier. Its average output voltage without a filter capacitor is 0.45 times the input AC voltage. One pulse occurs per cycle of AC input, as shown in Figure 8-4. Thus more filtering is required to provide a smooth DC output without ripple. For this reason, this circuit is limited to applications where the load is small, such as transmitter bias and high voltage cathode-ray tube power supplies. The peak reverse voltage (PRV) that the rectifier can tolerate varies with the load in the half-wave circuit of Figure 8-4. The rectifier PRV need only be 1.4 times AC input voltage with a heavy load, however, with a capacitive load where little current is drawn, a PRV as high as 2.8 times AC input voltage may be required.

AMATEUR CIRCUITS IN DETAIL

PRV Rating = 1.4 AC Voltage

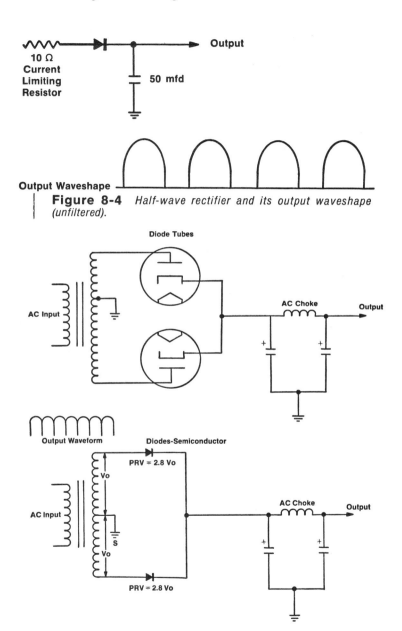

Figure 8-4 *Half-wave rectifier and its output waveshape (unfiltered).*

Figure 8-5 *Full-wave rectifier used with center-tapped power transformer. Peak reverse voltage (PRV) for each rectifier must be 2.8 the voltage output (VO) of half the transformer secondary.*

In Figure 8-5, a full-wave rectifier circuit is shown in both solid-state and vacuum tube versions. The outputs of the two rectifiers are combined so that rectification occurs during both halves of the AC cycle. A transformer with a center topped secondary is needed for this circuit. Average output voltage is 0.9 times the RMS voltage input. With a load under 50 MA when using a capacitor input filter, 1.4 times the RMS input voltage can be obtained. The PRV required for each rectifier is 2.8 times one-half the transformer secondary voltage.

The full-wave rectifier of Figure 8-5 is satisfactory when the power supply transformer secondary has a center tap. However, if it has none, then where is the negative power supply lead to come from? If we place four rectifiers together to form a bridge circuit, then the center tap on the secondary is no longer needed. Figure 8-6 shows the full-wave bridge rectifier configuration. On one-half of the cycle, current flows through the top right rectifier and the bottom left. When the cycle reverses itself, then current flows through the remaining two rectifiers. The diode's cathode (BAR) is always the positive side, while the anode (arrow) is always the negative side. The max output voltage into a resistive load or a properly designed choke-input filter is 0.9 times the RMS voltage delivered by the transformer secondary; with a capacitive input filter and a very light load, the output voltage is 1.4 times the secondary RMS voltage. The PRV (peak reverse voltage) per rectifier is 1.4 times the secondary RMS voltage. Each rectifier in this circuit must have a minimum load-current rating of one-half the total load current.

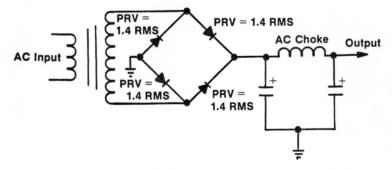

Figure 8-6 *Full-wave bridge rectifier. This circuit is used when the power supply transformer does not have a center tap.*

SURGE PROTECTION

When the power supply switch is first thrown, the large filter capacitors across the rectifiers have not yet charged. They look like a short circuit to the rectifiers. Surge protectors in the form of series resistors usually are installed to protect the rectifiers until the capacitors have charged. Such resistors before or after each rectifier may be from 10 to 25 ohms. This small resistance will protect the rectifiers for a few milliseconds, but will not affect the output voltage once the filter capacitors are charged. Another problem is transient voltages on the AC line. These short spikes come from nearby motors, power tools, air conditioners, oil burners, etc. Remember that the reactance of a capacitor decreases as the frequency increases. If we place a .01 mfd capacitor across the power transformer primary, it will have a high resistance (285,000 ohms) at 60 Hertz. However, at the higher frequency of the transient voltage spike, the reactance will be low so that the same capacitor will look like a very low resistance, shorting most of the harmful transient.

Transient protection can also be achieved by placing a .01 mfd capacitor across the rectifier itself. This prevents the temporary high voltage transient from damaging the rectifier. When two rectifiers are placed in series, as in Figure 8-7, the inverse voltage does not divide equally. To equalize the reverse voltage drop, equalizing resistors are placed across each rectifier. These may be 380,000 to 470,000 ohms with a one watt power rating.

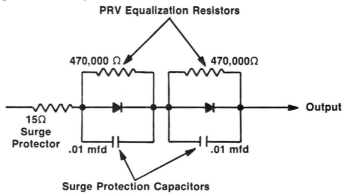

Figure 8-7 *Protecting diode rectifiers against surges and transients.*

OSCILLATORS

In the last chapter, we learned about the quartz crystal and the piezoelectric effect. The crystal determines the frequency of an oscillator or RF generator. Also, it is a delicate device that can handle very little power without fracturing, which is caused by heating or RF crystal current. Crystal drive or excitation should always be the minimum necessary for proper operation.

To design an extremely stable crystal oscillator, its output voltage should be small (with a light load) and the crystal drive level should be extremely low. A tube or transistor can be the active element in the crystal oscillator. In Figure 8-8a, the 6C4 triode vacuum tube serves as the active element in the Pierce oscillator. The crystal feeds energy from the plate (in-phase) back to the grid to maintain oscillations. The RF choke in the plate lead keeps the RF at the plate from being shorted out through the power supply. The .01 capacitor keeps DC off the crystal. The .001 capacitor prevents DC from appearing in the output circuit. In Figure 8-8b, the dual-gate MOSFET (metal-oxide semiconductor, field-effect transistor) becomes the active oscillator element. This oscillator will operate at extremely high frequencies—its power supply may be a 9 volt radio battery. Gate 1 is provided with a fixed positive bias through the 150K resistor. Gate 2 is connected to its own bias resistor and the frequency determining crystal. The source is grounded, and L and C are tuned to the oscillator frequency or an harmonic of it. RF output is taken from the drain. By inserting a telegraph key in the source lead, this circuit could be made into a shirt-pocket 100 milliwatt telegraph transmitter.

A tenth of a watt may not be sufficient excitation to drive high power vacuum tube circuits. We might need more than five watts of power to accomplish this. The 12BY7 in Figure 8-8c is capable of delivering this amount of RF power. In this electron-coupled oscillator, crystal current is coupled to the cathode where feedback takes place. There are no tuned circuits, so that only the crystal need be changed in order to change the oscillator's frequency. Let's suppose there are 200 volts on the 12BY7 plate, and this circuit draws 70 milliamperes. To calculate the DC power input, we would multiply the plate current by the plate voltage: 200 v x .07 amperes = 14 watts. If

AMATEUR CIRCUITS IN DETAIL

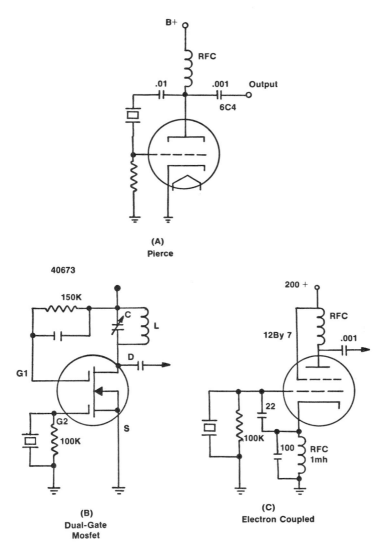

Figure 8-8 *Various types of oscillator circuits commonly used in amateur circuits.*

the DC input power is 14 watts and the oscillator's efficiency is 40 percent, then the RF output is 5.6 watts. This is sufficient to drive the power amplifier of a 50 watt transmitter. In all of these oscillators, we had to change crystals for the oscillator to function on a different frequency. We certainly could not tune a knob so that the oscillator would operate on any given frequency.

In Figure 8-9, we see the two basic types of variable frequency oscillators. In each type of VFO, a tuned circuit, instead of a crystal, determines the oscillator's frequency. By switching taps on L or adding additional capacitors to C, the tuned circuit may be made to tune over more than one amateur band. For example, it can tune 7 to 7.4 MHz (40 meters) as is. When a 150 mmfd silver mica capacitor is placed across C, it then can tune the entire 80 meter amateur band, or 3.5 to 4 MHz. The distinguishing feature of the Hartley oscillator is that feedback to the cathode is accomplished inductively via a tap on L. The distinguishing feature of the Colpitts oscillator is that feedback is accomplished through two capacitors (C1 and C2) acting as a voltage divider. To remember that the Colpitts is capacitive coupled, think of the letter C for Colpitts and capacitor. If the plate voltage should be kept extremely constant, if the tuned circuits (L and C) are well designed and kept at a constant temperature, the stability of the Hartley or Colpitts oscillators will be adequate for most low frequency applications up to 7 MHz. However, oscillators with fundamental frequencies above 7 MHz should be crystal controlled so that stability can be obtained.

SYNTHESIZERS

Up to 30 MHz, it is possible to operate a transmitter using a harmonic of a VFO (Hartley or Colpitts) which operates on a fundamental frequency of approximately 7 MHz. For example, a VFO operating at 7.1 MHz would yield a 4th harmonic in the 10 meter band, at 28.4 MHz. Does this mean that all transmitters above 10 meters must use only crystal control? Until recently this was true. As 2 meters became more popular, this became a serious problem. Each time it was necessary to operate a new frequency, you had to invest $10 in a new set of crystals and wait for their delivery. Worse than that, if your transmitter only held 6 sets of crystals, then you could only operate on 6 channels.

The advent of the synthesizer has now solved this problem. Synthesizers permit the flexibility of the VFO (operation on the

AMATEUR CIRCUITS IN DETAIL

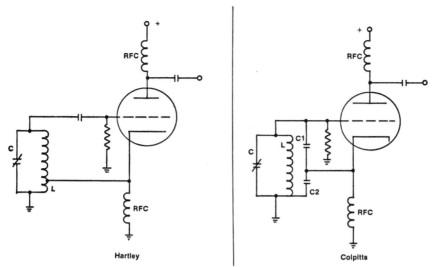

Figure 8-9 *Feedback in the Hartley oscillator is through an inductor. Feedback in the Colpitts oscillator is through two capacitors (C1 & C2).*

frequency of your choice) at the very high frequencies. Essentially, a synthesizer is a voltage tuned VFO operating at a very high frequency. Some of the oscillator's output is fed into a prescaler. This is an integrated circuit that divides (scales down) a frequency many times. The output of this circuit is a phase (or frequency) difference which is fed into a phase-locked loop. This loop then retunes the VFO, or maintains a constant phase. In other words, the prescaler and phase-locked loop combination constantly monitor the VFO output and correct drift or deviation in its output frequency. By adding integrated circuit technology to amateur radio circuits, we have a high frequency, stable VFO that was unheard of only a few years ago.

As of 1978 several manufacturers were producing 2 meter synthesized transmitter-receivers. These rigs used the prescaler, phase-locked loop techniques. They could tune to any frequency over the entire 2 meter band, 144 to 148 MHz, without crystals. Yet their stability was equal to transmitter-receivers using crystals.

Regardless of whether a crystal or a tuned circuit (as in the VFO) determines the oscillator's frequency, the basic principle is always the same. A method of feedback from the output circuit to the input circuit is necessary to sustain oscillations. In the Hartley, Colpitts, or other type of VFO, an LC circuit determines the oscillator's resonant frequency. In the crystal oscillator it is the crystal. The resonant frequency of an oscillator is the rate of alternating (AC) pulses per second. A 4 MHz oscillator is generating 4 million pulses a second. This is the oscillator's operating frequency.

RF AMPLIFIERS

An audio amplifier that amplifies voice or music must amplify a wide range of frequencies; that is, it is a broad-band amplifier. For this reason, it cannot have tuned circuits. On the contrary, the transformers in the amplifier should be as "flat" as possible over a broad frequency range. In other words, audio transformers should not favor any frequency in the audio spectrum, if good fidelity is to be achieved.

When working with high frequencies in the RF range, an amplifier must handle a single frequency or a narrow range of frequencies. For example, the RF amplifier in a receiver amplifies the weak signal coming in from the antenna. Its bandpass (range of frequencies it will pass) should be as narrow as possible to discriminate against radio stations on nearby frequencies which might cause interference. Thus tuned circuits in the input and output discriminate against signals on adjacent channels. Figure 8-10 illustrates a typical RF receiver amplifier. The 40673 is a tiny, inexpensive, dual-gate MOSFET which will work up to 500 MHz. L1 and L2 (input and output tuned circuits) utilize a variable inductor. Inside the coil is a threaded core which is adjustable with a screwdriver. This is slug tuning. The third circuit, L and C, match the antenna impedance to a low impedance tap on L1 so that the extremely high impedance gate (G1) is not loaded down and is protected against any electrical charges on the antenna. Usually, in an RF amplifier of this type, its high gain would cause a slight amount of feedback from the drain to the gate. Even this slight feedback would allow oscillation to occur. Neutralization to cancel the effect of this feedback would then be required. However, the MOSFET has a negligible capacitance from drain to gate, so that

AMATEUR CIRCUITS IN DETAIL

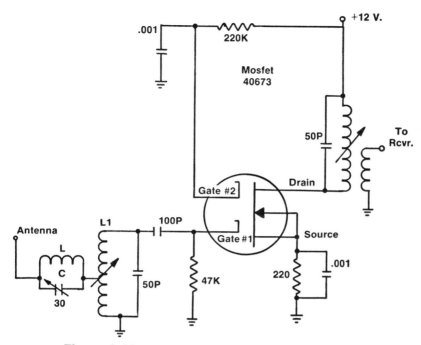

Figure 8-10 *A tiny, solid-state RF amplifier that will increase the sensitivity of any receiver. The inductance of L1, L2, and L depends on the frequency range.*

almost no feedback can occur. Thus in this case, no neutralization is required. Being Class A, this amplifier reproduces the weak signal faithfully, but cannot deliver power to the output.

When we build a transmitter, we will want to send power into the antenna so that we can reach great distances. Remember, the pebble only makes tiny ripples when thrown into the water, but the stone makes large ripples. In amateur radio, we are allowed to operate an RF power amplifier at 1000 watts DC input power. In AM (amplitude modulation) or CW (telegraphy), the fidelity of this amplifier is relatively unimportant. However, it would be advisable to operate this amplifier at the highest possible efficiency. This means the amplifier tube or transistor would dissipate less heat, and the RF output power would be higher. For these reasons, the amplifier could be operated in a mode where the plate (collector) current flows over a small part of the operating cycle. This is called a Class C amplifier, as shown in Figure 8-11.

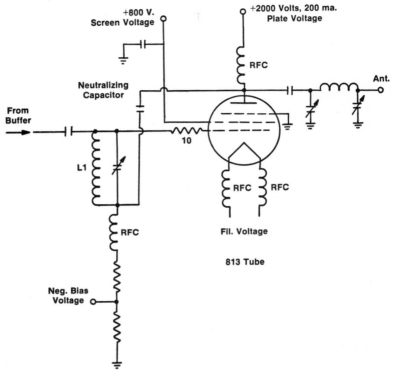

Figure 8-11 *This RF power amplifier can handle 500 watts. Two 813s in parallel will handle 1000 watts.*

The 813 transmitting tube can handle 500 watts of DC input power. In the illustration, the plate voltage is 2000 volts, the plate current is 200 milliamperes (0.2 amperes). Thus, the DC input power (E × I = P) is 400 watts. If the efficiency of this circuit were 70 percent, and the transmission line and antenna impedances properly matched, then 280 watts of RF energy would flow in the antenna. It is possible for this relatively high efficiency to be obtained because the amplifier functions in the Class C mode, on a single frequency. Single frequency operation results from the high Q (high ratio of reactance to resistance) inductors, L1 and L2. Variable capacitors across both inductors enable both parallel-resonant circuits to be tuned to the exact operating frequency. This is the frequency that is generated in the crystal, Hartley, or Colpitts oscillator (Figures 8-8 and 8-9). Since L2 must dissipate a great deal of power, it is composed of heavy wire wound on narrow polystyrene insulating struts. This is an air-core inductor.

The variable capacitor, C1, is connected from the 813 plate back to the bottom end of L1. This is the neutralizing capacitor. Inter-electrode capacitance inside the 813 (from plate to grid) can feed some output energy back into the grid circuit. This would cause the 813 to become a high powered oscillator. Its output would be fed into the antenna. Instead of other amateurs hearing your voice or telegraph signal, they would hear a raw, unstable signal which would swish across many frequencies causing a great deal of interference. The FCC would soon send you a citation or notice of a violation of good operating procedure. After several such notices, your amateur license would be revoked. The neutralizing capacitor (C1) intentionally feeds back some RF energy from the plate circuit. This RF energy and the grid-plate inter-electrode capacitance are going to opposite ends of L1. This means they are 180 degrees out of phase with each other. When C1 is tuned so that both feedback paths are equal, then one path cancels out (neutralizes) the other—the RF amplifier cannot oscillate.

The input of this amplifier receives power from a buffer stage which in turn receives its power from the oscillator. The buffer raises the low power from the oscillator stage to a value of excitation suitable to drive the 813 PA. As this tube is a high gain pentode, it has a high power gain sensitivity. That is, high power output is possible with very low power drive to the tube's input. For example, 500 watts of DC power can be applied to the 813 plate, with only 5 watts of RF power from the buffer driving the 813 grid. Two of these tubes can be wired in parallel; each element in one tube is connected to the similar element in the second tube. With this arrangement, two 813s are capable of operating at the maximum legal power—1000 watts.

RF TANKS AND ANTENNA COUPLING

Having just learned how RF amplifiers can handle a large amount of power, we now come face to face with another problem. What is the best way to transfer power from the RF amplifier to the antenna? Another problem is how to minimize the amount of harmonics from the transmitter that is radiated by the antenna. A major factor in both of these considerations is the Q of the tank circuit. Tank circuit Q determines how well the tank circuit (in the amplifier plate circuit) maintains its efficiency while it is delivering most of its energy to an antenna. In other words, what happens to the tank circuit when it is heavily loaded?

A Q of 10 to 20 is considered optimum. A lower Q results in less efficient operation as the amplifier begins to act like a broad-band amplifier instead of a single frequency amplifier. Thus harmonic radiation increases sharply, and coupling into the antenna (remember the antenna and ground constitute a tuned circuit) becomes much more difficult. Q is determined by the LC ratio of the tank circuit and the load resistance of the RF amplifier tube. The tube's load resistance is related to the ratio of DC plate voltage to DC plate current. In a Class C amplifier, the plate load resistance is calculated by:

$$RL \text{ (plate load resistance)} = \frac{\text{plate volts}}{2 \times \text{plate current}}$$

Thus the load resistance of our 400 watt 813 RF amplifier is calculated as:

$$\frac{2000 \text{ volts}}{2 \times 0.2 \text{ amperes}} = 5000 \text{ ohms}$$

This tells us that the output circuit (plate circuit) of the 813 tube (with 2000 volts at 0.2 amperes) looks like 5000 ohms to any tank circuit connected to it. This means that a 5000 ohm circuit connected to the 813 plate would result in a perfect impedance match, hence a perfect transfer of power.

Knowing this, we can then calculate the LC tank ratio that will give us 5000 ohms to match the 813 plate circuit. The higher the C in the tank, the higher the Q; however, high C is sometimes difficult to achieve because large variable capacitors are far more expensive than inductors.

In Table 8-1, we can see the tank capacitance required for a Q of 10, which is the minimum desirable Q for a loaded tank circuit. Doubling the capacitance doubles the Q.

AMATEUR CIRCUITS IN DETAIL

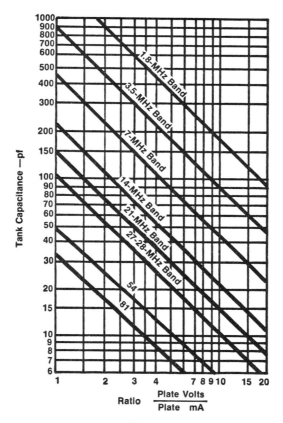

Table 8-1   *Courtesy ARRL*

Having matched the amplifier's tank (LC) circuit to the vacuum tube (or transistor), we can now proceed to matching the tank to the antenna transmission line. It is becoming customary in amateur radio to use low impedance transmission line or coaxial cable. Its losses are low, and its shielding prevents radiation on the way to the antenna. This is especially important whenever a medium or high power transmitter is involved. Coaxial cable in impedance is usually 50 to 72 ohms; refer to the coaxial cable table for particular types of coax.

The plate tank circuit of our 813 amplifier had a 5000 ohm impedance. This is very high compared to 50 ohms. Thus it is necessary to find a way to convert a high impedance into a low impedance to feed the antenna transmission line. One method of doing this is inductive-link coupling. A link is a small number of turns of heavy wire placed close to the cold end of the tank coil as shown in Figure 8-12. The optimum link is one whose self-inductance has a reactance at the operating frequency which is equal to the characteristic impedance, Z_o, of the antenna transmission line. While the tank circuit is a resonant circuit (tuned to the operating frequency), the link is not resonant. Thus its presence may cause some detuning of the tank circuit.

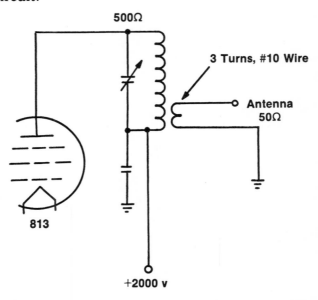

Figure 8-12 *Link coupling. One method of coupling a 5000 ohm tank impedance to a 50 ohm antenna transmission line impedance.*

At very low frequencies, the coupling of the link may be inadequate to transfer enough power from the tank into the transmission line. Another problem that develops is that a transmission line is not always flat. That is, an SWR (standing wave ratio) on the transmission line may raise its impedance at the transmitter-transmission line junction. In either of these instances, link coupling will create some problems in effective coupling. In addition, if we wish to feed a 300 ohm or higher transmission line, link coupling is useless. To gain additional flexibility in coupling a tank to higher impedance, and to

AMATEUR CIRCUITS IN DETAIL

overcome problems with a link at low frequencies, another coupling technique is often used.

PI-OUTPUT TANKS

It is possible to design a tank circuit that will allow coupling to a transmission line with an impedance anywhere between 20 and 6000 ohms. A pi-tuning network can also couple a transmitter's output directly into an antenna, without transmission line. However, this should never be used for voice transmission because the RF power in the radio shack will cause feedback. Figure 8-13 shows the pi-network circuit. The radio frequency choke (RFC) allows the DC to return to the power supply, but does not pass the radio frequency energy. C1, the coupling capacitor, isolates the DC voltage, passing only RF to the pi-network. C2 and C3 tune the inductor to the operating frequency. Think of C2 and C3 as capacitors in series with a ground between them. They are both part of a seesaw

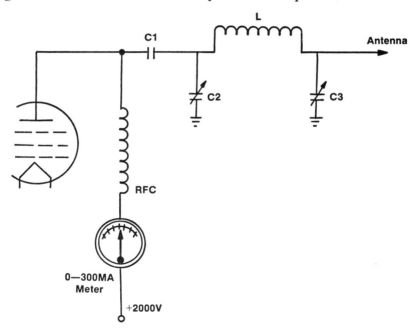

Figure 8-13 *The pi-network.*

configuration. When the capacitance of C2 is increased, then the capacitance of C3 must be decreased to compensate. In this case, the impedance at the antenna terminal would be as high as the reactance of C3 would be low. To match the pi-network into a low transmission line impedance, C2 is decreased and C3 increased to compensate. Now the reactance of C3 is higher. This higher reactance creates a low impedance patch. Thus the pi output is now matched to a low impedance output. The important fact is that both C2 and C3 constitute the variable capacitors of the LC tank circuit.

To determine when C2 and C3 are both tuned to make the LC circuit resonant at the operating frequency, several methods can be used. A plate current meter in series with the RF choke will show a slight dip at resonance. Also, a small neon bulb placed close to the tank circuit will glow when these capacitors are tuned to resonance. If the transmission line impedance is properly matched to the antenna impedance at its junction, the antenna will absorb all the power available from the pi-network. Thus if 1000 watts of DC power is applied to an RF amplifier whose efficiency is 72 percent, then 720 watts of RF energy will be radiated by the antenna.

9

Typical Questions on the Novice FCC Exam

The Novice Class license is the best way to get started in amateur radio. You will be able to operate a transmitter with power up to 250 watts. The Morse code part of the FCC exam only calls for you to send and receive at 5 words per minute. Using the techniques in the last chapter, you will have no trouble passing the code part of the test. The written part of the Novice exam covers only the most elementary aspects of amateur regulations and electronic theory. You may operate telegraphy in these bands: 3700-3750 kHz, 7100-7150 kHz, 21.1-21.2 MHz, and 28.1-28.2 MHz. This chapter covers sample questions and answers that you will find in the Novice Class FCC exam. If you can answer 75 percent of the sample questions, you will pass this part of the Novice exam.

RULES AND REGULATIONS

1. Q: Describe Part 97 of the FCC Rules. Also, what are the maximum penalties for violating Part 97?
   A: Part 97 of the FCC Rules governs the Amateur Radio Service. Its violation may bring a fine of $500 for each day in which the violation occurred.

2. Q: How would you describe the Amateur Radio Service?
   A: It is a radio communication service of self-training, inter-communication, and technical investigation carried on by amateur radio operators.

3. Q: Define an amateur radio station. Define an amateur radio operator.
   A: An amateur radio station is licensed in the Amateur Radio Service and includes the needed equipment at a particular location used for amateur radio communication. An amateur radio operator has an interest in radio technique without a personal goal, without a financial interest. The operator holds a valid FCC license to operate amateur radio stations.

4. Q: FCC Rules promote skills in which two phases of the art to improve the Amateur Radio Service?
   A: In the communications and technical phases of the art.

5. Q: May an amateur transmitting station be operated without an FCC license?
   A: No. An amateur radio transmitter may not be operated without an FCC license.

6. Q: May a Novice Class license be renewed? For how long is it valid?
   A: It is not renewable but a new examination may be taken and a new license obtained 12 months after expiration of any FCC amateur license. The term of the Novice license is two years.

7. Q: Where should you keep your amateur radio operator license? Your amateur radio station license?
   A: Your operator license should be kept in your personal possession or may be posted in a conspicuous place in the room where you keep your radio equipment. Your station license should be posted in the room where the equipment is kept or in your personal possession.

8. Q: Describe which persons may hold an amateur radio station license.
   A: Only a licensed amateur radio operator. Also a military recreation station license may be issued to a person not licensed as an amateur radio operator, who is in charge of a military recreation station not operated by the U.S. Government but located in approved public quarters.

THE NOVICE FCC EXAM

9. Q: Define a control operator. Who may the control operator of an amateur radio station be?
   A: An amateur radio station license may designate a control operator to be responsible for transmissions from that station. The control operator could be the station licensee or another radio operator.

10. Q: Which person has responsibility for the proper operation of an amateur radio station?
    A: The licensee is always responsible for proper operation. Also, the control operator is responsible for the station's operation.

11. Q: The Novice Class license gives you what frequency privileges?
    A: 80 Meters—3700-750 kHz   15 Meters—21.1-21.2 MHz
       40 Meters—7100-7150 kHz   10 Meters—28.1-28.2 MHz

12. Q: How would you describe an amateur radio station's log? How long should you keep it and what information must it contain?
    A: It is a record of the station's activity. The log should be kept one year. It should contain a copy of the station license, the station's location, dates at each location, dates and times of all visiting operators, with the signature and primary station call sign of any control operator. Also, an indication of messages sent and received for others—third parties—including the names of people speaking over the mike. Third-party traffic information may be recorded to be put into written form later. Additional information that may be recorded in your log is call signs, date and times you talk to another station, the frequency used, equipment changes, transmitter problems that have occurred. A subsequent FCC violation notice or any inquiry from another station might make the additional log information necessary.

13. Q: Your Novice Class license allows you to use what maximum transmitter power?
    A: Transmitter power input may not exceed 250 watts.

14. Q: The Novice Class license allows you to use what types of emission?
    A: Only A1 emission or telegraphy may be used. International Morse Code may be sent by keying the transmitter on and off. This is pure continuous wave called CW.

RADIO PHENOMENA

15. Q: What is the speed of radio waves in free space (meters per second)?
    A: Radio waves travel at approximately 300 million meters per second.

16. Q: How should the transmitting frequency of an amateur radio station be measured?
    A: Various devices may be used. A frequency counter, frequency meter, wave meter, or a receiver of known accuracy which has already been checked against a known station such as WWV operated by the National Bureau of Standards.

17. Q: Describe the propagation phenomena that permits radio waves to be transmitted great distances. Which frequency bands would you use for greatest distance during daytime? Also at night?
    A: Layers of ionized gasses in the ionosphere bend radio signals (sky waves) back toward the earth. These signals then return to earth a good distance from their originating point. This skip distance varies with frequency and the height of the ionized cloud, which changes with the time of day, time of year, and the 11 and 22 year solar sunspot cycle. During the daytime hours you would use 20, 15, and 10 meters to reach long distances. At night you would use 80 and 40 meters (longer wavelengths) to make long distance contacts.

18. Q: How is frequency translated into wavelength? List the wavelengths that the Novice Class license allows you to operate on.
    A: Frequency (in hertz) multiplied by wavelengths in meters equals 300 million. Thus, wavelength is

THE NOVICE FCC EXAM   171

inversely proportional to frequency. Or, the longer the wavelength, the lower the frequency. When wavelength in meters is known, find frequency (f) by:

$$f = \frac{300 \text{ million}}{\text{wavelength}}$$

to find wavelength, use the following:

$$\text{wavelength} = \frac{300 \text{ million}}{\text{frequency}}$$

Example: The wavelength of a signal whose frequency is 7200 kHz is:

$$\frac{300{,}000{,}000}{7{,}200{,}000} = \frac{300{,}000}{7{,}200} = 41.6 \text{ meters}$$

Novice frequencies translated into wavelengths become: 3700-3750 kHz, or 80 meters; 7100-7150 kHz, or 40 meters; 21.1-21.2 MHz, or 15 meters; 28.1-28.2 MHz, or 10 meters.

OPERATING PROCEDURE

19. Q: Describe the RST reporting system. For example, what would RST579 mean?
    A: RST stands for readability, signal strength, and tone. Readability ranges from 1 (unreadable) up to 5 (perfectly readable). Signal strength is rated from 1 (barely perceptible) up to 9 (very strong). And tone is rated from extremely rough A/C note Morse code tone of 1 up to 9, which is perfect tone, with no trace of ripple or modulation. RST579 would be readability perfect, signal moderately strong, and Morse code tone crystal clear and smooth.

20. Q: When operating telegraphy, describe what these letters would mean: AR, SK, CQ, DE, K?
    A: AR signals the end of a message or round of conversation; SK indicates the end of a group of messages or of the contact (the line over the letters means they are run together and sent as one character); CQ means calling any station; DE means 'from' or 'this

is' (used before the call sign); K means "go ahead, your turn to transmit."

21. Q: When you get your license, how should you pick your transmitting frequency? What special precautions should be followed when selecting a transmitting frequency near one end of the authorized amateur band?

    A: Review question 17 on how to select the proper band. To pick the best frequency within that band, listen for a clear spot to avoid interfering with other stations. If you pick a frequency near the edge of a band, consider the percentage of error inherent in the device you used to indicate your frequency. Add to this one-half the bandwidth of your signal. A voice signal's bandwidth is its two sidebands. A telegraph signal's bandwidth is determined by the speed of your Morse code. Bandwidth in hertz equals 4 × speed in words per minute. Thus when you send code at 9 words per minute, your transmitter's bandwidth is (9 × 4) 36 hertz.

22. Q: Why are Q signals used? Define the following Q signals: QRM, QRU, QRZ, QTH, QSL, QRS.

    A: The International Telecommunication Union has adopted Q signals so that complete sentences could be transmitted in a brief period of time.
    QRM—Interference is making your signal difficult to receive.
    QRU—I have no messages for you.
    QRZ—Standby, or wait before you transmit as another station is calling you.
    QTH—My latitude or longitude is, or my location is ...
    QSL—I recognize your transmission.
    QRS—Please send more slowly (words per minute).
    A question mark after the Q signal turns it into a question.

23. Q: Specify the ingredients that make up high quality telegraphy (A-1) transmissions.

A: There should be no clicks as the key opens and closes. Harmonics and bandwidth of the signal should be at a minimum. The A-1 telegraphy signal should be a pure note (DC) and constant in pitch without chirping as it is keyed on and off.

24. Q: Describe A-1 emission.
A: A-1 is used in the amateur rules to indicate telegraphy by keying the transmitter on and off without an audio modulating frequency.

THEORY OF ELECTRICITY

25. Q: Describe the following types of electrical phenomena: 1. direct current 2. alternating current
What is the function of a rectifier?
A: DC current flows through conductors in only one direction. Batteries are a source of DC. AC changes periodically from negative to positive and back again. Current flows in one direction, then another. The rate of change is known as the frequency which is expressed in hertz (cycles) per second, kHz (kilohertz), or MHz (megahertz) per second. A rectifier is a device that changes AC into DC current. A filter of capacitors follows the rectifier and eliminates any ripple or 60 hertz in the rectified output.

26. Q: Define current, electrical power, EMF. How are each of these parameters measured?
A: Current is the rate of flow of electrons through a conductor. Expressed as amperes or milliamperes, electrical power is the ability of electricity to do work, expressed in watts or kilowatts. EMF, or electromotive force (E) is the electrical pressure, measured in volts. The relationship between voltage (E), current (I), and resistance or ohms (R) is expressed by Ohm's Law which is $I = E/R$, $R = E/I$, and $E = I \times R$.

27. Q: Distinguish RF from AF frequency.
A: RF or radio frequency is very high, while AF, or audio frequency, is low and audible.

28. Q: Define hertz, kilohertz, megahertz.
    A: Hertz is one AC cycle (positive and negative reversal) per second, kilohertz is a thousand cycles per second, and megahertz is a million cycles per second.

29. Q: Define resistance, inductance, and capacitance, and give the units of measurement for each.
    A: Any obstacle to the flow of electrons through a conductor is resistance, measured in ohms or megohms. Inductance is energy stored in a magnetic field such as a coil of wire. Capacitance is energy stored in an electrostatic field, such as two conductors separated by an insulating dielectric. This constitutes a condenser or capacitor. Capacitance is measured in farads, microfarads (millionth of a farad), or picofarad (millionth of a millionth). Inductance is measured in henrys, millihenrys (thousandths), and microhenrys (millionths).

30. Q: How is a second or third harmonic related to its fundamental frequency?
    A: A harmonic is always an integral multiple of its fundamental frequency. Thus the second harmonic is twice the fundamental, while the third harmonic is three times the fundamental frequency.

CIRCUITS

31. Q: Draw a circuit having an ammeter to measure current, a voltmeter to measure voltage, and a resistive load.

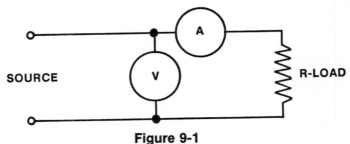

Figure 9-1

32. Q: Using the diagram you have drawn, describe how you would measure the value of the resistive load and the power consumed by this load.

A: Ohm's Law, R × E/I, tells us that the resistance equals the voltage across it divided by current flowing through it. The voltmeter and ammeter would thus give us the values needed to calculate the resistance of the resistive load. The power consumed by this load equals the voltage across it times the current flowing through it, P = E × I. Again, the voltmeter and the ammeter give us the values needed to calculate the power consumed by the load.

33. Q: Using the same circuit again, what would be the value of the battery voltage if 2 amperes flowed through a resistive load of 45 ohms?
A: In Ohm's Law (E = I × R or current in amperes × resistance in ohms), 2A × 45 ohms equals 90 volts, which is the battery voltage.

34. Q: Draw a transmitter Power Amplifier including coupling capacitors, bypass capacitors, meters, pi-network circuit, triode tube, RF chokes, and neutralizing capacitor.

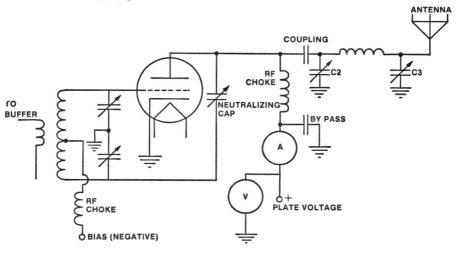

Figure 9-2

35. Q: How would you tune up a neutralized power amplifier?
A: Start with the buffer providing signal power to the amplifier. Tune C2 so that the plate current dips or

reaches minimum, then tune C3 for a peak. Repeat this procedure until the power (voltage × current) is close to the desired wattage level. To neutralize this amplifier, switch off the plate voltage but let the buffer signal drive the power amplifier. The ammeter will show a slight indication due to capacitive coupling from the grid to plate (anode) of the amplifier. Tune the neutralizing capacitor until this slight plate current is zero, or has been neutralized.

ELECTRONIC COMPONENTS

36. Q: You should know the schematic symbols of a choke, resistor, capacitor, inductor, transformer. Draw these.
    A: choke resistor capacitor inductor transformer

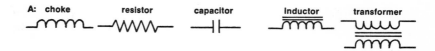

37. Answer true or false:
    Q: 1. An insulator blocks the flow of electrons and has high resistance. Examples are mica, quartz, glass, ceramics, and rubber.
    2. Conductors have low resistance and allow electrons to flow. Silver, copper, aluminum are examples.
    3. Semiconductors have qualities halfway between insulators and conductors. Examples are silicon and germanium. Semiconductors are used to make transistors and diodes.
    A: 1. True   2. True   3. True

38. Q: Draw the schematic symbol of a diode, transistor, triode vacuum tube.

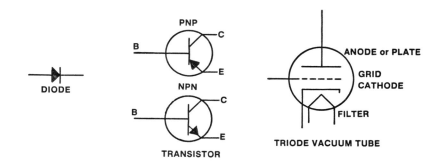

Figure 9-3a   Figure 9-3b   Figure 9-3c

39. Q: Describe the theory of a half-wave antenna. Give the approximate lengths in feet for half-wave antennas for the bands set aside for Novice Class use.
    A: Half-wave antenna is one-half wave long at the frequency for which it is cut. The formula for determining its length is:

    $$\text{length in feet} = \frac{468}{\text{frequency in MHz}}$$

    From this formula, the antenna halfwave length for the various Novice bands can be calculated; 80 meter band—126 feet; 40 meters—66 feet; 15 meters—22 feet; 10 meters—17 feet.

40. Q: Describe a dipole antenna.
    A: A dipole is one-half wavelength long. Transmitter output is applied at a low impedance point at the center of the antenna.

41. Q: List two advantages of a multiband antenna. What are its limitations?

A: These antennas save space. They allow operation on several bands. Their limitations are that they favor undesirable harmonics and subharmonics, increasing the possibility of interfering with other stations. A solution to this problem is to connect a multiband antenna to the transmitter through a matching inductor (transmatch), which is tuned to accept only the desired frequency.

42. Q: What is the name for the line which connects a transmitter to its antenna? Name and describe three types of these lines.
A: This line is called a transmission line. One type is the coaxial line. An insulated conductor is encased in a shield or wire braid. Another type is the open wire line. Two separated wires are joined by insulators with an appearance like a ladder. A third type of line is twin lead, with two wires molded inside a plastic tubular jacket.

43. Q: Your radio station will have a receive/transmit switch, ground rod, antenna, transmission line, telegraph key, transmitter, speaker, receiver. Draw a *block* diagram of your radio station and include all these components.

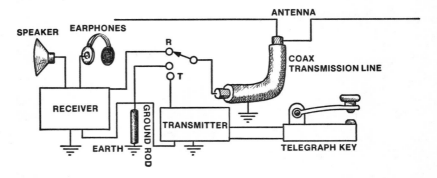

44. Q: Each component you have drawn in the last question has a different function. What is it?
A: The transmit/receive switch connects the transmitter to the antenna when you are sending, and connects the receiver to the antenna when you receive. A ground

THE NOVICE FCC EXAM   179

rod is a metal rod driven into the ground. It drains unwanted electrical charges away from the equipment, reducing the danger of shock. It can also be connected to the antenna when the equipment is not in use, thus allowing lightning to bypass the equipment. A lightning arrester between the ground rod and the antenna will do the same thing. The receiver converts signals from the air waves into audible sounds which come from the speaker or earphones. The transmitter generates a radio signal on a particular frequency or wavelength. This signal is turned on and off in the form of dits and dahs (Morse code) by the telegraph key. The coaxial transmission line carries the transmitter energy to the antenna.

45. Q: How can the danger of electric shock be reduced at an amateur radio station?
    A: Use an external ground such as a water pipe or ground rod. Use power transformers, not AC/DC power supplies. Use a three-conductor power line with a ground wire (green). Keep antenna wires and the transmission line far away from the electric power line. Use electronic components and wires heavy enough to carry currents and voltage going through them. Use bleeder resistors across power supplies to discharge voltage when the equipment is turned off. Use an interlock switch to disable power supplies when you are working on the equipment.

46. Q: What is a device that matches the impedance of the transmitter to the antenna or the transmission line? Name two of its advantages.
    A: A transmatch. It ensures maximum energy transfer to the transmission line. It attenuates undesired harmonics which could be radiated causing interference.

47. Q: What is the best way for an amateur to determine the quality of transmission from his/her radio station?
    A: The most commonly used method is to have a local amateur station give you a report on the quality of your

voice or code transmission. You might also connect an oscilloscope to your transmitter to determine if its signal is "clean."

48. Q: Would your Novice Class transmitter with the following voltages and current be legal? Explain your answer.
    Power Amplifier Plate Volts—1,100 volts
    Power Amplifier Plate Current—240 milliamperes
    Filament Voltage—6.3 volts
    Filament Current—2 amperes
    Driving Power—1 watt
    Screen Grid power—1.2 watts

    A: Your transmitter would be illegal. The Novice Class license allows you to operate a transmitter with maximum power input of 250 watts. To determine the power input, do not include the filament voltage or current. The plate power is 1,100 volts times 240 milliamperes (or .240 amperes). Thus 1100 × .240 equals 264.00 watts. Now add the driving power (1 watt) and the Screen Grid Power (1.2 watts) to the Plate power. Thus, the total Power Amplifier power input is 266.2 watts. As the maximum power allowed is 250 watts, this transmitter is illegal.

49. Q: Name the best ways to determine that your transmitter is radiating inside an authorized frequency band.

    A: You can measure the transmitting frequency with a frequency meter, a wave meter, or a receiver of known accuracy. To ensure the accuracy of a receiver, calibrate its frequency dial against calibration stations of a known frequency. Examples of these stations are WWVH, CHU, and WWV at 2.5, 5, 10, 15, and 20 megahertz. WWV also gives data every hour on shortwave propagation conditions, the weather, and sunspots.

ARE YOU READY FOR THE EXAM?

There are several electronic dictionaries which can be extremely helpful as an aid in studying the material found in

THE NOVICE FCC EXAM 181

this chapter. The ARRL publishes several basic electronics primers including *The Radio Amateur's Handbook*. Amateur radio retail stores sell a variety of electronics and amateur radio magazines which will help you to become more familiar with the theoretical material found in this chapter.

When you have sufficiently mastered both questions and answers included here, then you will be ready for the written part of the FCC exam. However, you are not yet prepared for the Morse code part of the exam. The next chapter will prepare you for that part of the FCC exam.

10

How to Benefit from the Amazing Navy Code Study Course

WHY CODE?

We have learned how an amateur license will permit us to operate a teletype set, amateur TV, or talk to other amateurs via the ionosphere, repeaters, or satellite. In order to pass any amateur radio test, it is first necessary to be able to send and receive the International Morse Code. The speed necessary to pass each of the amateur tests is 5 words per minute for the Novice test, 13 words per minute for the General Class test, and 20 words per minute for the Extra Class test.

Using the techniques specified in this chapter, you should be able to pass the Novice test after two weeks of study. This goal is based on a study session of one hour every day for two weeks. Why is the FCC so insistent on prospective amateurs knowing Morse code? As we learned in Chapter 3, Morse code was the mode of communication long before voice transmission was ever devised. Long after voice transmitters were designed, ships in distress would still use telegraphy to get their message through. The air waves are often loaded with static and all types of atmospheric noise. Weak voice transmissions seldom are received during these conditions. Another reason for using telegraph is that it requires less spectrum space than voice transmission.

In addition, it is often necessary to operate a transmitter from emergency battery power. Not only does telegraphy allow a low power battery transmitter to be heard, but it also consumes a fraction of the power used by voice transmitters. Remember the simple telegraph transmitter described in Chapter 6. It consumes power only when the telegraph key is down. Thus, battery power would last a long time. That is why aircraft pilots who are forced down in the sea always have tiny beacon transmitters that operate on Morse code. These transmitters automatically send SOS, the internationally recognized 'MAYDAY' or distress signal. Hundreds of amateur operators over the years have received acclaim for picking up and reporting SOS signals. Telegraphy is also used when power lines are down during floods, fires, earthquakes, and hurricanes. A lesser known use for telegraphy is for the transmission of secret messages. You might receive the message BUUBDL on your receiver. If you knew the key to break this code (in this case each letter has been moved one up in the alphabet), then you could decipher this message— ATTACK. When practicing Morse code with your friends, see how many more code keys you can originate.

THE NAVY'S METHOD

When World War II broke out, the U.S. Navy found itself with a gigantic problem. It had to train thousands of telegraph operators in a hurry when the mothball fleet was put into service. The Navy called upon renowned psychologists to help it with its problem. It was these psychologists who were able to design a course that enabled raw recruits to become skilled telegraphers within a short time. This course was based on three principles: 1) Visual learning methods are just as important as auditory methods; 2) Recite the letters, numbers, and punctuation marks of Morse code exactly the way they sound; and 3) Games make learning much easier.

Thus a visual learning technique was devised. Cut a large piece of cardboard into a small card or square for each Morse code symbol to be described later. On one side of the card write

THE AMAZING NAVY CODE STUDY COURSE

the letter or number; on the other side write the Morse code equivalent. The object of the learning game is to guess the dot-dash when the letter is face up. The psychologists insisted on something else. They said dot-dash does not sound like the actual sound of telegraph. Dit-dah is closer to the actual sound. The psychologists then replaced the dot-dash with the dit-dah.

In carrying out their belief that learning games make everything much easier, the psychologists divided the letters of the alphabet into groups that could be made into 'fun' stories. The first group is the easier one. This consists of the following letters:

```
E .         T -         A .-        N -.
I ..        M --        W .--       J .---
S ...       O ---       U ..-       D -..
H ....                  V ...-      B -...
```

In this half of the alphabet, EISH are all dits, TMO all dahs, AWUV is a dit-dah combination, and NJDB is a dah-dit combination. Cut the cardboard cards for these letters.

Now go on to the remainder of the alphabet.

```
K -.-       G --.       R .-.       F ..-.
C -.-.      Z --..      X -..-      L .-..
Y -.--
Q --.-
```

The last half of the alphabet is combinations of dits and dahs. Cut out cardboard cards for this half of the alphabet.

Now cut out cardboard squares for the numbers and punctuation marks as follows:

```
1 .---              6 -....
2 ..---             7 --...
3 ...--             8 ---..
4 ....-             9 ----.
5 .....             0 -----
```

Period .-.-.- Comma --..-- Question mark ..--..
Error .... (long series of dits) Double Dash -...-
Colon ---... Semicolon -.-.-. Parenthesis -.--.-
Fraction Bar -..-. Wait .-... End of message .-.-.
Invitation to transmit -.- (k) End of Work ...-.-

A MORSE CODE OSCILLATOR

With an oscillator or buzzer and a key you will develop your fist (sending) while training your ear (receiving). Figure 10-1 is a schematic diagram of a transistor code oscillator. Figure 10-2 is a diagram of a common doorbell buzzer which can be purchased in any hardware store. The two lantern type dry cells yielding 3 volts will last over a year even if you practice every day. Your local electronics store sells several types of code practice oscillators. I would recommend the AMECO, Model

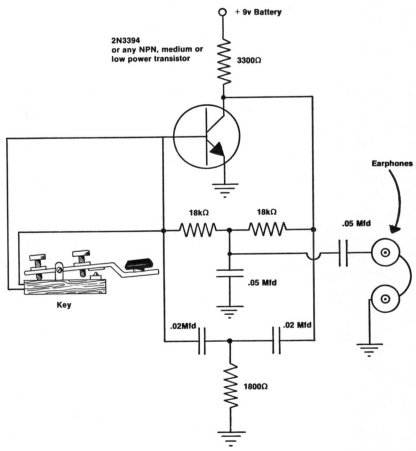

Figure 10-1 *The twin-T audio frequency transistor code oscillator. This circuit eliminates the need for a bulky audio transformer.*

CPS, or even the AMECO code buzzer, Model B-1. The AMECO MODEL KB-1 is a self-contained code practice set having a buzzer, a key, and a battery holder. If you decide to buy a key, they are now priced as low as $1.25 and you must mount them on a table or a wooden board.

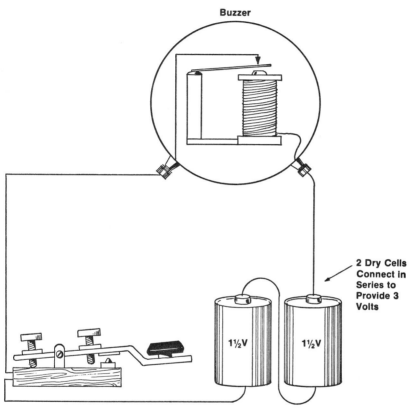

Figure 10-2 *The common doorbell buzzer is connected to a battery and key to make a code practice buzzer.*

RULES FOR CODE PRACTICE

The following rules, like any, can be altered according to your particular preference. However, they have come into existence as a result of a hundred years of code practice training sessions. If you refer back to these fundamental rules whenever a problem develops, you will probably discover what you are doing wrong.

1. Never study when tired or not feeling well.
2. Limit each study session to less than 2 hours a day.
3. Increase your code speed very slowly.
4. Be sure to take a break every 15 minutes.
5. Spend more time with the difficult letters and punctuation marks.
6. Mix numbers and punctuation marks in with sentences.

Several of these general rules have to do with study when you are fatigued. Learning Morse code requires an alert mind and a great deal of concentration. Therefore, studying when you are fatigued or when there are distractions (such as TV) is a waste of time. One word on the FCC exam consists of 5 letters (with numbers mixed together). In order to pass the code part of the Novice exam, you must send and receive one word every 12 seconds (5 words per minute). After you have been practicing for a week or two, use the second hand of your watch to check your speed.

RULES FOR SENDING

After you have played the guessing game several times with the cardboard squares, you will be ready to use the telegraph key. Following these rules will enable you to learn to transmit more easily. Remember that listening to your own sending is the best way to develop your ear, or receiving capability.

1. The key should be mounted to a large board or the table itself. There should be enough room for your elbow to rest on the table (20 inches). Refer to Figure 10-3.
2. It is best to hold the key in your strongest hand unless you cannot write with the other hand.
3. Sit erect. The table should be about 30 inches high.
4. Sit comfortably, with your shoulders parallel to the table. Your legs should not be crossed, and your feet should be on the floor.

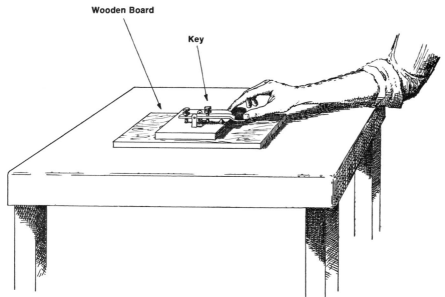

Figure 10-3 *Telegraph operator rests elbow on the table. The key is bolted to heavy wooden board.*

HOLDING THE KEY

Study Figure 10-4 for the best position for holding the key. High speed telegraphers all follow this procedure; it will make the sending much easier. The thumb is firmly pressed against the edge of the knob. The index (first) finger is curved slightly and touches the top rear of the knob. The second finger is also

Figure 10-4 *Different students may prefer different way to hold the key. However, this is the correct way to obtain best results.*

curved and placed against the rear of the knob opposite your thumb. The remaining fingers curve toward the palm and do not touch the knob.

ADJUSTING YOUR KEY

Adjust the two side screws in Figure 10-5 for free movement of the lever without side play occurring. Next, adjust the rear screw so that the contact points are 1/32 to 1/16 inch apart when the key is not depressed. Use the wider contact separation when you start to practice. Later, when your speed increases, you will want to make the contact gap closer. Adjust the spring tension screw so that the contacts separate as soon as the knob pressure has been released. When there is excessive spring tension, your wrist will quickly become fatigued. When there is too little spring tension, the dits will be uneven and choppy, and the spacing between dits and dahs will be poor.

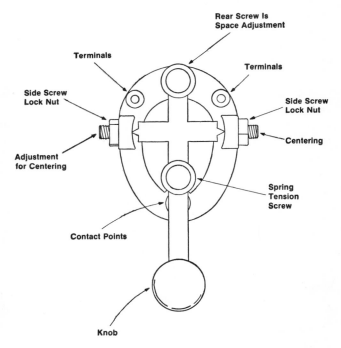

Figure 10-5 *Top view of telegraph key. The two wires are connected to the terminals. Space and spring adjustments can be changed after skill is developed.*

WARMING UP TO THE KEY

Once you are comfortable with the way you are holding the key, and assuming you have followed the rules, you are ready to begin to send. Connect the key to the code oscillator described in either Figure 10-1 or 10-2. A dah is much easier to send than a dit, so start with dits (dot). Send about 3 per second. Make sure your fingers do not leave the knob, and at the same time do not grasp it too tightly. Send about 8 dits in a three second period, listening to determine if the spacing between dits is even. Always keep your wrist flexible, allowing it to move loosely during sending. You will find that most of the work is done by your wrist.

The difference between dits and dahs, as well as the spacing must be mastered before going any further. The beginner might send very long dahs to make them easier to recognize.

EXERCISES

Remember the easier half of the alphabet. These are the letter groups; EISH, TMO, AWUV, NJDB. Refresh your memory regarding this half of the alphabet by picking these cardboard squares out and playing the guessing game with them several times. When you feel you are familiar with these letters, send the following story, without punctuation marks. This nonsense story contains only the half of the alphabet we are now studying.

> Joe and Jane saw someone down the avenue. It was a woman in the town, named Beth. She waved to them. The two visited Beth and ate. Jane had meat and a bun. Tea swished in Joe's mouth. His jaw moved high above his dish. Then Joe, Jane, and Beth went to the window to view the moon.

When you feel that you can send the easy half of the alphabet with confidence, you are ready for the remainder of the alphabet. Pick out the cardboard squares KCY, GZ, RX, FL. Play the guessing game several times with these squares. Now send the following nonsense story which contains these letters and some from the easy half of the alphabet. Omit the punctuation marks and numerals.

Xavier (the quick but lazy hunter) killed 4 geese for food once: geese were abundant then—not like today. He remembered another incident; that he killed 7 young quail in zero degree weather. Was Xavier brave, 6 years ago?

Now select the cardboard squares that represent numerals and punctuation marks. Play the guessing game several times using these squares. When you think you know punctuation marks and numerals, send the nonsense story about Xavier again. This time include numerals and punctuation marks. When this exercise becomes tiresome, try the nonsense story of Joe and Jane with punctuation marks to break the monotony. Another helpful exercise is sending all punctuation marks together, then all numerals together. You will notice the dit-dah pattern of 1-5 (an increasing dit) and the pattern of 6-0 (an increasing dah).

As a further aid in receiving, you might wish to purchase one of the many cassette code practice tapes on the market. Your local electronic distributor may carry them. Any of the amateur radio magazines usually carries advertisements for mail order firms that sell these practice tapes at a low price.

Another helpful aide is WIAW, the shortwave station operated by the ARRL. WIAW broadcasts Morse code at various speeds on a regular schedule. QST magazine will provide you with current broadcast schedules and frequencies.

KNOWING WHEN YOU ARE READY

The story about Xavier is the equivalent of 35 words. If you can send this story in seven minutes, then your sending speed is 5 words per minute. This is the requirement for the Novice FCC exam. It is true that only 75 percent of the code received or sent on the test must be correct to pass. However, most applicants are nervous when taking their test and make more errors than during practice. Therefore, a wise precaution is to be able to send and receive 7 words per minute for the Novice Code test. This means you should be able to send the story about Xavier in 5 minutes, and be able to receive it at the same speed. If you can do this, you are now ready to take your Novice FCC exam. Remember, if you fail the exam, you can take it again after a 30 day waiting period, so there is no need to be nervous. After passing your novice test, the FCC will issue you your own call letters. Then you can broadcast around the world with the transmitter described in Chapter 6, or with any of many commercial Novice band transmitters. Within a short time, this on-the-air experience will raise your code speed to 13 words per minute. By studying more advanced theory than the questions given in the last chapter, you will then be ready to pass the General Class amateur exam, which will entitle you to operate all amateur bands will full privileges.

Good luck.

11

How to Build Your Own Set

IN THE EARLY DAYS

Today any amateur can buy receivers and transmitting equipment from $30 to $3000. However, in the early days of amateur radio, hams always made their own equipment. In most cases, they even made their own components by winding transformers, coils, and using metal plates as condensers. Most of the large electronic corporations of today were founded by these early amateurs who became skilled in building electronic components themselves.

The enjoyment in those days was in putting parts together (or making your own parts), connecting the finished product to a homemade antenna, and sending telegraphy or talking to amateurs far away.

A HOMEMADE TRANSMITTER

When you begin your amateur radio hobby, you will want a sensitive receiver, with good selectivity, that covers all the amateur bands. A receiver can be purchased for anywhere from $50 to $600, and building one yourself is very difficult for the beginner. On the other hand, building your own transmitter will cost only $20. In addition, you will learn a great deal about transmitter theory, and wiring and soldering practice. The satisfaction of contacting distant stations with a homemade transmitter cannot be equaled by using a purchased, commercial transmitter. Before we can begin to think about building anything electronic, however, we must learn the basic rules of construction practice.

CONSTRUCTION PRACTICE

Figure 11-1 lists the transmitter components with their pictures. Collect all the components that will be required to build your Novice telegraph transmitter. Place each component in the position it will appear in the left side of the tube, and the other components (right of the dotted line) on the other side of the tube. This will prevent oscillation in the power amplifier. If the power transformer and power supply components in Figure 11-2 are mounted on the same chassis as the transmitter itself, make sure at least 3 or 4 inches separates the transmitter components from the power supply components. This is necessary to ensure that 60 Hz, AC hum from the power supply does not modulate the transmitter, causing a rough chirp (birdie) in its telegraphy tone.

Measure the space needed to mount all the components that have been collected. Add a few inches in both directions and buy an aluminum chassis having these dimensions, or as close to them as possible. The depth of the chassis is not critical. Mounting hardware is sold at all electronics distributors. When drilling large holes in the chassis, make sure it is clamped securely so that it cannot spin. A spinning chassis can be very dangerous.

A soldering gun is preferable to a soldering iron, which takes a long time to heat up. If a low wattage gun is used, it will take some time to solder the heavy leads of a power transformer. Clean the soldering gun tip at frequent intervals.

To cut a rectangular hole for the power transformer, cut a series of small holes along the lines that will ultimately be cut. An easier method is to drill a large round hole in one corner of the chassis. Then, with a hacksaw, saw the four corners into a square or rectangle, as in Figure 11-3. A triangular file will do this job too, but will take longer.

HOW TO BUILD YOUR OWN SET

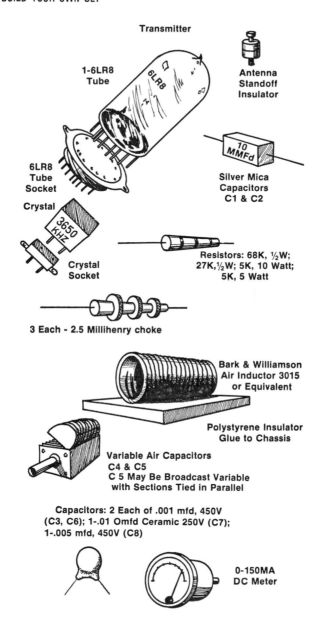

Figure 11-1 *Pictures of the components needed to build the one-tube globe trotter telegraph transmitter.*

HOW TO BUILD YOUR OWN SET

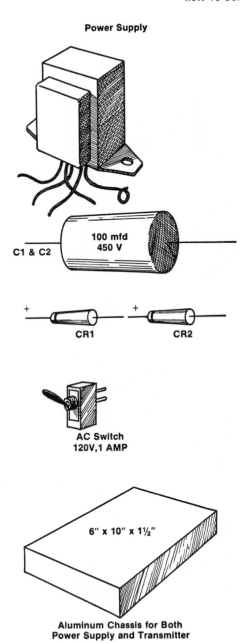

Figure 11-2

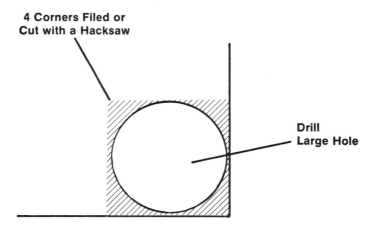

Figure 11-3 *The easiest way to cut a rectangular hole in a chassis. Small holes may be used instead of a single large hole.*

To determine the value of resistors and capacitors, refer to Table 11-1. The color coding of various transformers is also given in this table.

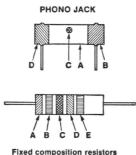

Fixed composition resistors

The colored areas have the following significance:
A—First significant figure of resistance in ohms.
B—Second significant figure.
C—Decimal multiplier.
D—Resistance tolerance in percent. If no color is shown the tolerance is ±20 percent.
E—Relative percent change in value per 1000 hours of operation; Brown, 1 percent; Red, 0.1 percent; Orange, .01 percent; Yellow, .001 percent.

Table 11-1

SOLDERING

After drilling the chassis, mount the tube socket, the crystal socket, the variable capacitors, the key jack, and the antenna jack. If you use the same chassis for the power supply, also mount the power transformer and the AC power switch. The meter may be mounted in a separate box or on the chassis using two small L brackets. Another possibility is that any ordinary multimeter with a 200 milliampere range be used instead of the meter. After the antenna has been built and the rig has been properly tuned up, the wires going to the multimeter can be taped together and the multimeter removed.

When soldering, make sure that the soldering gun tip is clean. First, hold the soldering gun firmly on the connection for a few seconds. Then apply the solder to the connection (not to the gun). Let the solder flow into the connection until the entire connection and wire is covered with solder. Never allow so much solder into the connection that a blob of solder surrounds the connection or solder runs off the connection. This is the easiest way to create a short circuit.

If the solder does not flow into the connection easily and appears to crack in places, you have a cold solder joint. This is the result of a dirty soldering gun tip, a defective tip, or simply not enough heat reaching the connection. Cold solder joints will cause numerous headaches. They may create a good connection at times and not at other times. Thus your rig will work only on an intermittent basis if at all. Before you make any soldered connection, make sure the right wire or component is attached to the correct numbered pin on the tube socket. A numeral on the tube socket base designates each pin. A magnifying glass will help you to read these numbers.

THE TRIODE-PENTODE

In the early days of radio, a vacuum tube was a triode, or a tetrode, or a pentode, the triode having one grid, the tetrode two, and the pentode three. As television circuits began to develop, it became fashionable to put two (and sometimes three) vacuum tubes within the same glass envelope. As you can imagine, this resulted in a sharp reduction in labor costs. The heart of the one tube telegraph transmitter we are going to build is a common TV tube with two tubes in one envelope. The 6LR8 is used in TV sets as both an oscillator and amplifier. The triode section is the vertical oscillator (sweeps the picture from top to bottom), and the pentode section is the vertical amplifier (amplifies the vertical sweep signal).

THE PROJECT

In our circuit, Figure 11-4, the triode becomes an electron-coupled oscillator with energy from the grid being fed back to the cathode by C1 and C2.

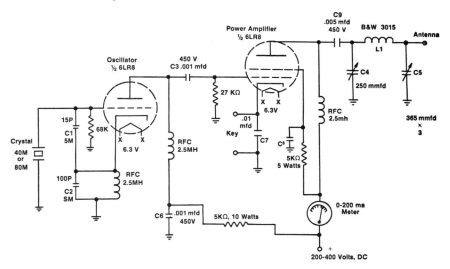

Figure 11-4 *Schematic diagram of the one-tube globe trotter telegraph transmitter. Figure 11-5 shows the power supply.*

The output of the oscillator is fed through C3, the mica coupling capacitor, to the grid of the power amplifier. The output of the power amplifier is fed to a pi-tuning tank circuit which consists of L1, C4, and C5. When C5 is completely meshed, it represents (365 mmfd × 3) 1095 mmfd. This is a very low impedance from the antenna post to ground. Thus very low impedance (50 ohm) coaxial cable could be connected to the antenna terminals. This coaxial transmission line is terminated in an antenna at a very low impedance (50 ohms) point.

A pi-tuning network can match a wide variety of output impedances, from 50 ohms up to 600 ohms. If the antenna terminals were connected to a 600 ohm transmission line, then less capacity (higher impedance) would be needed at C5. It would no longer be fully meshed. However, C5 and C4 both form the parallel resonant tank circuit along with L1. This means that as C5 is unmeshed (less capacity), the mesh of C4 must be increased (more capacity) to compensate.

If a B and W 3015 inductor cannot be obtained for L1, then 22 turns of No. 20 wire can be wound on a plastic form one inch in diameter. If only 40 meters (7 MHz) will be the operating band, L1 can be reduced to 19 turns. If the Hammarlund Mc250M cannot be obtained for C4, and a lower value variable capacitor must be substituted, then L1 will have be more than 22 turns. Capacitors C3, C6, and C7 are not critical, and capacitors close to these values can be removed from old TV sets. The same applies to resistors R1, R2, and R3. Make sure R3 is at least one watt. All other components can be secured from an electronic distributor.

THE POWER SUPPLY

The power supply for our one tube telegraph transmitter consists of a power transformer, two solid-state silicon rectifiers, two capacitors, and two resistors. Figure 11-5 is the schematic diagram.

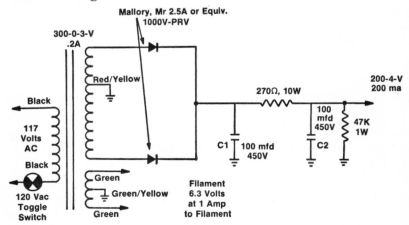

Figure 11-5 *Power supply for the one-tube transmitter. If the power transformer is taken from an old TV set, the 5 volt rectifier filament winding is not used (yellow leads).*

HOW TO BUILD YOUR OWN SET

The transformer, T1, may be any transformer with a center-tapped high voltage winding that will supply from 250 volts to 400 volts. The low voltage filament winding must be 6.3 VAC for the 6LR8 filament. It must be able to supply a one amp load. Almost any power transformer from an old TV set can be used for T1.

Any surplus electronic parts store will usually have a used power transformer that can be purchased for a few dollars. Amateurs often hold flea markets where used parts are sold. Vacuum tube power transformers are easy to locate at these flea markets.

The two filter capacitors can also be taken from old TV sets. Make sure the correct polarity is observed. The specified capacitance (100 mfd) is not critical, and any value down to 20 mfd might be used instead. The lower values, however, might result in a slight chirp or rough tone when sending Morse code.

The two rectifiers, CR1 and CR2, could also be taken from old TV sets. However, if they are not at least 1000 PRV when used with a 400 volt power transformer, they might blow up due to excessive voltage being placed across them.

In the full-wave rectifier circuit described in Figure 11-5, the PRV or P1V rating of each rectifier must be 2.8 times the voltage appearing across one-half the power transformer secondary. Assuming this is 350 volts, then the minimum rectifier PRV rating must be 980 volts. The resistor R1, and the capacitor C2, provide some more ripple reduction (filtering of AC) so that the Morse code note is clean and crisp.

The 47K resistor is a bleeder. It allows the filter capacitors to discharge so that they do not store energy when the set is off. This might create a dangerous situation when you decide to work on the power supply, even after it has been turned off.

TUNING UP

If you have a multimeter, test the power supply for the 6.3 volt, AC filament voltage, and the high voltage DC. The high voltage will read 50 (or more) volts higher than normal, without the transmitter connected to it. Connect the transmitter. Tune your shortwave receiver to the frequency of the crystal you are using. You should hear a very strong CW signal from the oscillator alone. As the telegraph key has not yet been depressed, the power amplifier is not on. If no oscillator signal is heard, recheck the wiring, see if the 6LR8 filament is lit, and change crystals. A very old crystal may have low output so that the oscillator will not oscillate. Make sure that C1 is 15 mmfd and C2 is 100 mmfd. These values can be critical.

Once you can hear the signal from the oscillator, wire a 60 watt incandescent light bulb from the antenna terminal to ground. Press the key down. Tune C5 for maximum capacity (fully meshed). Tune C4 until the incandescant light reaches maximum intensity. This should be close to the position on C4 where the 0.200 milliampere meter indicates a slight dip. The dip, however, may not be seen if the light bulb is absorbing all the power available from the power amplifier. Now remove the light bulb and connect your antenna. If you are using coaxial cable transmission line (low impedance), the meter should show a slight dip near the position of C4 where the light bulb lit the brightest. A neon bulb placed near L1 or C6 should also light the brightest where the slight dip of the meter is observed. If the dip is very sharp, then the antenna is not loading up; that is, the VSWR is high or very little energy is going into the antenna. In this case, work must be done on the antenna so that it matches the transmission line. The next chapter will cover an antenna that will tune up very well with this one-tube globe-trotter.

12

A Guide to Great Antennas You Can Easily Build

After you have passed the Novice test, you can operate CW (telegraphy) on approved portions of the 80, 40, 15, and 10 meter bands. If you have built the 50-watt globe-trotter explained in the previous chapter, only the lack of a good antenna is holding you back.

In deciding whether to operate on 80 or 40 meters, it would be wise to consider these facts. There is much more interference on 40 meters. This will be a definite handicap unless you have a very expensive receiver and a high power transmitter. As amateurs with the expensive equipment tend to be up on 40 meters, they are also the more seasoned amateurs. This means their code speed ranges anywhere from 10 to 20 words per minute. If we can assume that your code practice sessions have enabled you to go only as far as passing your Technician's test, then your code speed is probably only about 6 words per minute. In this case, contacts with the high speed operators on 40 meters will not be much fun. However, the operators on the Novice part of the 80 meter band will send and transmit anywhere from 6 to 13 words per minute. Most of them will keep you alert and help you to increase your speed to 13 wpm for the General Class license.

80 METER DOUBLET

In Chapter 5, the half-wave doublet was described as the most efficient of all the antennas. Let's start by building a half-wave doublet for 80 meters. You can use this with your globe-trotter transmitter to make contacts in many countries.

Our 80 meter doublet will be deisgned for 3.7 MHz. Figure 12-1 shows the length of a half-wave doublet for different frequencies from 80 meters down to 10 meters. This scale is based on the formula:

$$\text{length of half-wave antenna (feet)} = \frac{468}{\text{freq.(MHz)}}$$

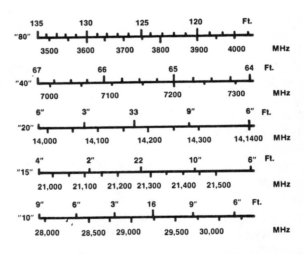

Figure 12-1 *Use this scale to determine the length of a half-wave antenna made of wire.*

The length of our 3.7 MHz half wave is then 126 feet Going back to Chapter 5, the half wave was described as really two one-quarter wave sections fed in the middle. Thus we actually have two wires, each being 63 feet long. They are separated by a small egg-type porcelain insulator. Two more porcelain insulators separate both ends from the wires attached

GREAT ANTENNAS YOU CAN EASILY BUILD

to trees or supporting poles. Figure 12-2 shows the 80 meter doublet mounted between two trees. It is center fed with coaxial cable (RG58/U) which is 52 ohms. The higher the antenna above ground, the lower its angle of radiation. The lower its angle of radiation, the better signal it will send to, and receive from, foreign countries thousands of miles away. Seventy or more feet above ground would be ideal. However, you probably will have to compromise, as trees this height are difficult to climb. Another possibility is to use 10 foot aluminum supporting poles which fit into one another. These are available

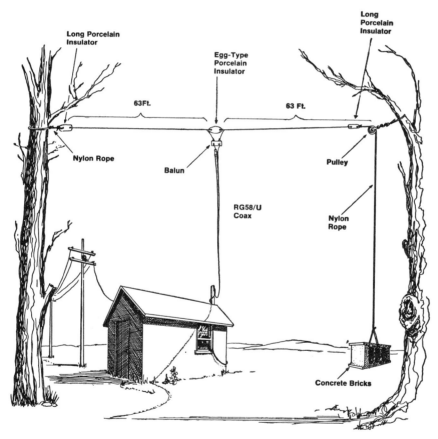

Figure 12-2 *This 80 meter doublet should be placed as high as possible for long distance communications. It can be used with the globe trotter in the last chapter.*

at TV supply and electronics distributors. If you fit seven of these together, it would be wise to use guy wires to hold these poles in place. Otherwise, the first heavy windstorm will blow them down.

WIRE SIZE

The preferred type of wire for antennas is copperweld. This is the highest purity copper wire with the lowest resistance. The wire gauge should be number 14 AWG. Number 12 might give slightly better performance, at much higher cost. Many more problems will be encountered in stretching it taut due to its increased weight. The lighter gauge number 16 might save some money and will be much easier to work with. However, the antenna's efficiency will be considerably less than that with number 14 wire.

Make sure that the two connections from the transmission line are soldered to the antenna, with electrician's tape protecting the soldered connections from the elements. A heavy soldering gun or iron must be used, since the antenna wire will quickly draw heat away from the connection.

THE BALUN

Coaxial cable can be connected directly to the 80 meter doublet without a balun. Figure 12-2 shows the balun connected in the transmission line to achieve perfect balance. However, if one of the quarter-wave sections is higher than the other, or more in the clear (not surrounded by trees), then this section should be connected to the cable's inner conductor (if a balun is not used). With the coaxial shield connected to the other quarter-wave section, much more energy will radiate from the section connected to the coax's inner conductor.

This is so because the balanced antenna is connected to coax which is an unbalanced transmission line. If you were lucky enough to get both quarter-wave sections high in the sky, you might wish to have both sections radiate an equal amount of energy. This will also increase the antenna's efficiency.

The balun is a device which accepts energy from an unbalanced transmission line. Its output is perfectly balanced so that both quarter-wave sections will receive an equal amount of the transmitter's energy. A close-up showing the balun inserted in the transmission line is shown in Figure 12-3. Baluns are now available in most retail outlets that sell amateur equipment.

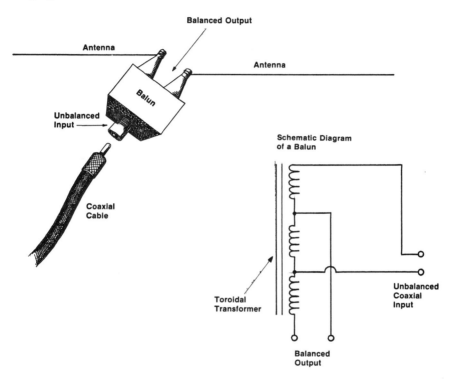

Figure 12-3 *How a balun is inserted between a transmission line and the antenna.*

TESTING THE ANTENNA

Before the 80 meter antenna is connected to a transmitter, it would be wise to test it. There are two basic antenna tests.

If you applied too much heat to the coaxial cable inner connector when you soldered it, and the cable was bent near the antenna connection, you may have melted the insulation surrounding the inner conductor of the coaxial cable. This would cause the inner conductor (where the insulation has melted) to touch the outer shield of the coax. This type of short is easy to detect. Connect an ohmmeter to the male plug at the transmitter end of the antenna's transmission line. If the resistance is infinite, there is no short. If the resistance is very low, you will have to do the job over.

The way to finally check out the antenna's performance is with a grid-dip meter. Wind a 15 turn link of number 28 wire below the coil of a grid-dip meter as shown in Figure 12-4. This link couples the antenna to the grid-dip meter. The dip meter, or dipper as they are now called, is simply an oscillator with external coils covering different frequencies. When another resonant circuit is close to the dipper's coil or an antenna coupled to it, the dipper's meter will dip when its calibrated dial is tuned to the resonant frequency of the coil or antenna being tested. Therefore, slowly tune the dipper's calibrated dial from 3 to 4 MHz. Watch closely for some kind of dip near 3.7 MHz. This will prove that the antenna is tuned to 3.7 MHz. A dipper is a worthwhile investment, for it can be used for many different tests. Several dippers in kit form are relatively inexpensive.

The dipper test is not absolutely necessary. If the antenna has passed the transmission line ohmmeter test, then we know it is not shorted. If you have been very careful in measuring the 63 foot length of each quarter-wave section, and the antenna is up high and in the clear, then it should perform very well.

THE SHORT 40 METER VERTICAL

The 80 meter doublet is a fine antenna if you have trees or supporting poles that are spaced 130 feet apart. However, if you are an apartment dweller or have a very limited amount of space available, the full length 80 meter doublet will pose a problem. There are two other alternatives. Buy a set of 80 meter loading coils which will enable your 80 meter doublet to be much shorter. Instructions with the loading coils will tell you how much wire to use with each coil.

Another alternative is to build a short 40 meter vertical. This antenna will have a very low angle of radiation which will favor long distance (DX) contacts. It will be omnidirectional. That is, it will radiate a signal equally in all directions.

GREAT ANTENNAS YOU CAN EASILY BUILD

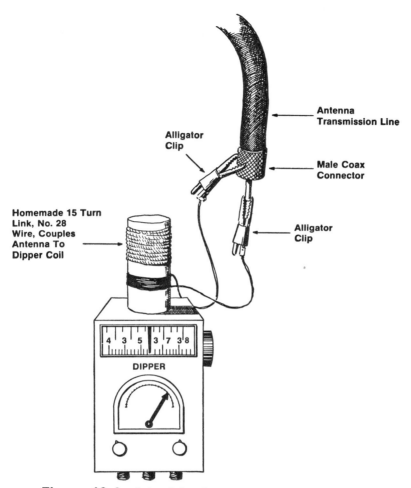

Figure 12-4 *Determining the resonant frequency of any antenna with a grid-dip meter, or dipper.*

LOADING COIL

The loading coil for this antenna can be homemade. It is wound on a 1/8 inch diameter solid plexiglass cast rod. Each end of the plexiglass rod is held in position with auto hose clamps. Figure 12-5 shows the loading coil attached to the coupling sections. This is the vertical quarter-wave element of the antenna. The wire size should be number 16 enamelled copperweld wire. Wind 65 turns on the plexiglass rod. Four large "L" brackets hold the bottom of the antenna to the flat roof of an apartment house. To attach this antenna to the top of a slanted roof, use two triangular wooden or plexiglass blocks nailed into the roof as shown in Figure 12-6. Shellac the wooden block to keep it moisture free. A small 365 mmfd variable capacitor tunes the antenna.

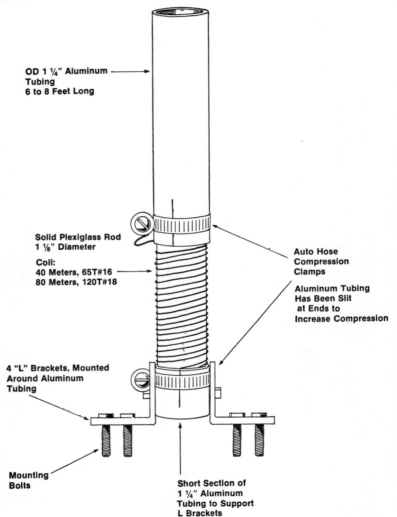

Figure 12-5 *Short 40 meter vertical antenna. By changing the loading coil to 120 turns of no. 18 wire, this becomes an 80 meter antenna.*

GREAT ANTENNAS YOU CAN EASILY BUILD   213

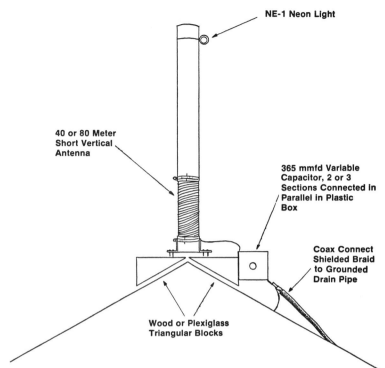

Figure 12-6 *A unique way to attach the 40 or 80 meter short vertical to the top of a slanted roof.*

To tune the antenna, start with the variable capacitor fully meshed. Connect the transmission line to a transmitter and tune the power amplifier milliameter for the resonant dip. Then, returning to the variable tuning capacitor, tune it until the transmitter power amplifier plate current has a very small dip. The RF power amplifier may have to be adjusted slightly as the antenna tuning capacitor is tuned. If the antenna is on the roof, far from the transmitter, it might help to run two insulated wires from the transmitter so that the RF power amplifier milliameter can be temporarily located near the antenna where it can be seen while tuning up. This tuning method is not preferred, as we are forced to run high voltage wires on the roof. This can be dangerous. The preferred method is to tune the antenna using a field strength meter. One can be purchased or built according to plans in Chapter 15. Figure 12-7 illustrates this tuning technique.

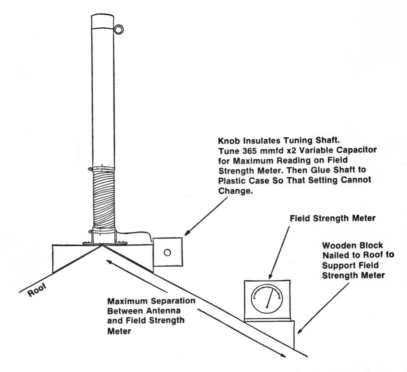

Figure 12-7 *The best method of tuning the 40 meter (or 80 meter) short vertical antenna.*

The antenna tuning capacitor should be placed in a sealed plastic box to protect the tuning capacitor from the elements. When the 40 meter vertical has been tuned properly, this 7 foot antenna will be equivalent electrically to a 32 foot wire. In other words, an actual quarter-wavelength in space.

Another way to tune the 40 meter (or 80 meter) vertical while on the roof is to tape a small neon bulb (type NE-1) to the top of the 72 inch aluminum tubing. On a cloudy day, or near dusk, tune the antenna capacitor while the transmitter is connected and turned on. When the antenna tuning capacitor is correctly tuned, the tiny neon bulb atop the antenna will glow brightly. Be very careful not to touch the antenna capacitor, except through an insulated knob, or you will receive an RF burn. The high frequency energy from the transmitter has ionized the neon so that it glows blue or orange. Needless to say, this method should never be used in wet or damp weather.

Holding a neon light at the end of a plastic rod near the end of an antenna is a convenient way to find out if energy is going into any antenna.

A SATELLITE SIGNAL-SQUIRTER

In 1978, the amateur satellite OSCAR 7, Mode B, had two days of the week reserved for QRP (low power) communications. The high sensitivity of the 432 MHz receiver aboard this satellite has baffled NASA engineers. They are at a complete loss to explain it. What this quirk means to the average amateur is that even 5 watts of RF power with a high gain antenna can trigger OSCAR 7. Even handie-talkies are available with 5 watts of power.

Simply by attaching a lightweight, balsa wood rod to stiff wire or aluminum welding rod elements, you can make a high gain OSCAR signal-squirter. Bolt the 6 foot balsa rod to the top of a handie-talkie. Figure 12-8 shows the 432 MHz, 11 element Yagi antenna. The elements are made of 3/32 inch or 1/8 inch diameter wire or welding rod.

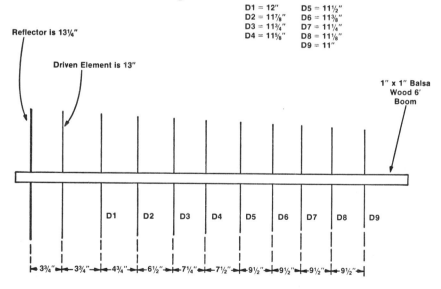

Figure 12-8 *The high gain OSCAR signal-squirter. With this 11 element Yagi, a low power 432 MHz transmitter can trigger OSCAR.*

The elements of this beam antenna can also be made out of number 10 electrician's wire, which is available in any hardware store. This copper wire has a higher electrical conductivity than aluminum and is thus slightly more efficient. Strip the wire's heavy insulation off with a knife, then use a cheesesaw to cut the wire to the correct length. Make certain the wire is perfectly straight. Holes are drilled into the 1 inch by 1 inch square balsa wood boom where the elements intersect the boom. Each element is inserted into its hole, then glued into place with epoxy cement. The longest element (the reflector) is 13-1/4 inches long, next is the driven element (energy is fed to this element), and the remaining nine elements are directors. They direct the transmitted energy away, in that direction, and they also direct the received signal from that direction. In other words, the signal is squirted out toward the shortest elements, just as a flashlight squirts light away from its reflector.

Although the antenna is 6 feet long, it weighs only a small fraction of a pound. This is made possible by the lightweight balsa wood boom which is mounted on a handie-talkie, the top of which is pointed to the satellite's elevation over the horizon. Thus satellite tracking is manual, and costly and bulky antenna tracking equipment is completely eliminated.

In Figure 12-9, we are looking at a front view of the driven element. Like all of the directors and the reflector, it is inserted through a snug hole in the 1 inch by 1 inch wooden boom. The element is centered, then glued. The delta match connects the tuning stub to the dipole at the point of best impedance match; in this case, 3 inches apart. Figure 12-9 shows the half wavelength balun. The balun is made out of the same material as the transmission line: RG8 foam dielectric. As can be seen in the data on coaxial cable, this type of coax has relatively low loss at 432 MHz. The balun allows the unbalanced transmission line to feed a balanced dipole. It must be one-half wavelength long. One-half wavelength in coax is shorter than a half wavelength wire, because the foam insulating material slows the wave down. This is the velocity factor. Thus the wave travels a shorter distance. F/492 gives us a half wavelength wire; F/492 × 0.8 (the velocity factor of RG8) gives us half wavelength in RG8 coaxial cable. Thus 0.91 feet is the length of the balun—11

GREAT ANTENNAS YOU CAN EASILY BUILD

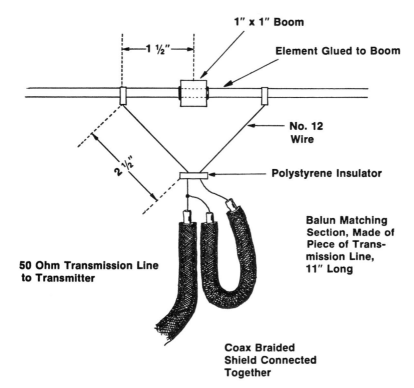

Figure 12-9 *A front view of the Yagi's driven element. The balun matches the unbalanced driven element.*

inches. The transmission line braided shield connects to the balun's braided shield. One leg of the delta match connects to the inner conductor of the transmission line and one end of the balun. The other delta leg goes to the inner conductor of the other end of the balun.

TWO-METER ANTENNAS

As soon as you receive your Technician's Class license, you will probably wish to operate on 2 meters or 144 MHz. Almost 3000 repeater stations in the U.S. make this band very enjoyable. Chapter 14 describes how repeater operation works.

There are several companies which manufacture efficient 2-meter antenna. Simple mobile antennas or complex co-linear arrays can be purchased inexpensively. Therefore, there has not been a project in this chapter describing the construction of a 2-meter antenna.

13

Preparing for the General and Technician's Exam

TECHNICIAN'S LICENSE

Having received your Novice license, having built the globe-trotter in Chapter 11, and having made many enjoyable telegraphy contacts, you are probably jumping at the chance to operate all the amateur bands using voice. The Technician's license will allow you to use voice on 6 meters (50.1-54 MHz), 2 meters (145-148 MHz), and all the higher frequencies. The Technician class code test is the same as the Novice class—5 words per minute. So having passed the Novice test, you should have no trouble with that. However, the written part of the Technician's test is more complex than the Novice test. The written part of the Technician's test is covered in this chapter. It is the same as the written part of the General Class Exam.

THE GENERAL CLASS EXAM

The Technician's license allows you to use voice on the high frequencies, but you still are prohibited from using voice on the low frequency bands. To use voice on all bands with complete privileges (except a very small segment of frequencies reserved for the Advanced Class licensee), you must get a General license. The General and Technician's test are the same, except that the candidate for General must pass a 13 word per minute telegraphy test. Both licenses have a 5 year term and are renewable upon application. Applicants for license renewal must show proof of two hours operating time over the past three months or five hours operating time over the last twelve months.

The required passing grade for both exams is 74 percent. The exam consists of 50 multiple choice questions. Sample questions follow. If you study these and review the sample questions in Chapter 9 on the Novice exam, you should have no trouble passing this part of the test.

SAMPLE QUESTIONS

RULES AND REGULATIONS

1. What is third party traffic? Which types are not allowed?

   This is traffic sent under the supervision of a control operator. The traffic is sent on behalf of a third party—not the two stations in contact. The following communication is prohibited:

   A. International third party traffic, except with consenting countries.

   B. Where compensation (material or intangible) is given to a third party, station licensee, control operator, or other person.

   C. Business communication which means communication which facilitates the regular commercial affairs of any party.

2. Define fixed operation, portable operation, mobile operation.

   Fixed operation is from the location shown on the station license. Portable operation is from a specific location, but not that of the station license. Mobile operation is conducted from a moving or stationary vehicle.

3. What kind of transmissions can take place between U.S. amateurs and foreign amateurs?

When permitted, transmissions must be in plain language. They shall be of a technical nature relating to tests and personal remarks. Amateurs may not engage in third party traffic with foreign countries.

4. List three types of stations amateurs may communicate with.
   A. Foreign stations, except where amateur radio is not allowed.
   B. Stations in other services and those of the U.S. Government for civil defense, emergencies, or testing.
   C. Any station authorized by the FCC to communicate with amateurs.

5. What rules must be followed during third party communications?

   A control operator must be present and must monitor and supervise communications to see that they follow the rules. Names of persons other than a control operator using the station for amateur communication shall be entered in the log. Third party international traffic is never allowed, except under special arrangement.

6. Who is responsible for proper operation of an amateur station?

   The licensee. Each control operator is also responsible. The signature of the control operator on duty and the station's call sign (if other than the station licensee) must be entered in the log.

7. Name the types of one-way Amateur transmissions that are allowed, also what types are prohibited?

   Transmissions are allowed for measurements, observations of propagation conditions, radio control, and other experiments. Emergencies and emergency drills. Information bulletins concerning amateur radio. Round table networks where many stations take part. Code practice transmissions. However, transmissions are prohibited that are intended to be received directly by the public or using relay stations as an intermediary.

8. Whose possession must the log be in?

   In the possession of the station licensee.

9. Can programs or signals from nonamateur stations be retransmitted by amateurs?

   NO. Amateur stations may not be used for this purpose.

10. How frequently and in what manner must an amateur identify his/her station with its call sign?

    At the beginning and end of each transmission, and more often than every 10 minutes. The call sign also must be given at the end of each transmission where a group of stations are in a round-table network.

11. Where a control operator is other than the station licensee, what are the operator privileges? How would the station be identified?

    The station must be operated according to the privileges of the class of amateur license held by the control operator. These privileges may exceed those of the station licensee if proper identification procedures are followed. When the control operator is not the station licensee, the station identification is the call sign assigned to that station. When the operator's class of license exceeds the privileges of the station licensee, the station's call sign must be followed by the operator's call sign (i.e. W2TCC/WA21PR).

12. Define a control point. Which amateur stations must have a control point?

    The operating position of an amateur station where the control operator function is performed is the control point. There must be a control operator at every authorized control point.

13. Where must notice of operation away from the authorized location of an amateur station be sent? When is it required?

Notice is sent to Engineer in Charge of the radio district of the intended operation. Outside the U.S., notice is sent to District Engineer having jurisdiction over the authorized location on the license. Within the U.S., advance notice is required for portable operation over 15 days. Outside the U.S., notice is required for any portable operation, regardless of its length.

14. Which of the following may be transmitted: music, secret codes and ciphers, obscenity, profanity, or unidentified communication or signals?

    None of these may be transmitted.

15. What are the rules for stations used for remote control of craft and vehicles?

    Remote control transmitters with power less than one watt may be operated, provided an identification card (or durable plate) indicates the call sign and licensee's name, and is attached to the transmitter.

PROPAGATION AND OPERATION

16. Can variations in the ionosphere's density (daily and yearly with respect to the 11 year sunspot cycle), ionospheric absorption, and noise influence radio propagation?

    Yes, all these factors influence propagation.

17. Name propagation influences on the VHF amateur bands.

    VHF influences are E layer changes and sometimes F layer changes. Also, aurora displays, meteor trails, and tropospheric bending due to temperature changes.

18. Can bending of skywaves back to earth produce communications?

    Yes, long distance communications can take place.

19. At VHF bands, is ionospheric bending common?

    No, it is relatively rare.

20. What do the following have in common?
    A. Using VHF bands for short distance communication.
    B. Using minimum power for communication.
    C. Using the minimum bandwidth.
    D. The use of directional antennas.
    E. Listening before transmitting.
    F. Using a dummy antenna for testing a transmitter.
    G. Using CW break-in, VOX (voice operated transmission), and short transmissions.
    H. Monitoring your transmission to detect distortion or poor telegraph quality.

    All of the above (A-H) are examples of good amateur operating procedure.

21. How are types of emission classified in the rules governing amateur radio?

    Type
    A0 RF carrier; not modulated; not interrupted
    A1 Telegraphy (conventional type) on RF carrier
    A2 Telegraphy by audio tone AM modulation
    A3 AM telephony; also Single Sideband
    A4 Facsimile
    A5 Television

    Type
    F0 RF carrier, not modulated; not interrupted
    F1 Carrier shift telegraphy
    F2 Audiofrequency shift telegraphy
    F3 Frequency or phase modulated telephony
    F4 FM facsimile
    F5 FM television
    P Pulse emissions

22. To transmit voice, what range of frequencies will allow good intelligibility?

    200 to 3000 Hz.

23. Should AM telephony be modulated over 100 percent?

    No, that would produce spurious radiation.

THE GENERAL AND TECHNICIAN'S EXAM 225

24. Can FM occur below 144 MHz when amplitude modulation is used?

   No.

25. Explain 100 percent amplitude modulation.

   The instantaneous amplitude and polarity of the audio signal changes the amplitude of the RF carrier from zero to double its normal level. This is called the RF envelope and exists because the RF carrier frequency is many times that of the audio modulating frequency.

26. Explain frequency and phase modulation.

   In FM, the audio varies the bandwidth of the RF carrier. In phase modulation, the audio varies the RF carrier bandwidth by changing the phase of current in the modulating stage. There is also a frequency change which depends on its speed and the amount of the phase shift. Both frequency and phase modulation may be detected in the same way.

27. What is narrow band FM or F3?

   NBFM is where the frequency deviation is limited to 5kHz. This deviation is commonly used on the VHF frequencies and contrasts with wide-band FM used by broadcast stations of 50 to 75 kHz.

28. What is another term for the bandwidth in which 99 percent of the total transmitted signal takes place?

   Occupied bandwidth.

29. In AM transmission, how is the average power input determined?

   Average power is taken over several cycles of identical modulation. The average power with single tone AM modulation is 1.5 times the unmodulated power. In this case the average power would be the DC input power multiplied by 1.5.

30. In single sideband modulation, how can the PEP (peak envelope power) be determined?

PEP occurs at modulation peaks. To find its value in watts, square the RMS voltage at RF envelope peaks, and divide the result by the impedance value across which the voltage appears. Using a dummy load and an oscilloscope, the RF peak value is taken with the oscilloscope connected across the dummy load. A single tone test signal allows the true voltage to be taken. A wattmeter will also give the PEP output when the single tone test signal is used. To find the PEP input power, multiply the plate or collector voltage by plate or collector current.

31. How is radio and TV interference eliminated?

Electrical devices can cause interference, commonly called RFI (radio frequency interference). Amateur transmitters often create TVI or interference with television sets. TVI is cured by installing a high-pass filter at the TV set, cleaning the TV antenna connections, or a wavetrap. Also, a low-pass filter can be added to the transmitter to suppress harmonic radiation. Sometimes shielding the transmitter is helpful. Relocating the transmitting antenna further away can be helpful if everything else fails.

32. What is the PEP-to-average-power ratio in single sideband (SSB) or A3 transmission?

The band-pass characteristics of the audio and filter systems along with the operator's voice determines the PEP-to-average-power ratio. While average SSB power may be only 75 watts, PEP power may be 150 or 225 watts—a ratio of 2 or 3 to 1.

33. Define the signal-to-distortion ratio in an SSB transmitter.

The S/D ratio is the ratio of the peak carrier amplitude and the third and fifth order (intermodulation) distortion products. A two-tone signal is fed to the transmitter and the intermodulation level is observed on a spectrum analyzer. The frequencies of the IMD products are found by multiplying the frequency of one tone by two and subtracting the frequency of the other one; the result is the third order IMD product. Fifth order products are found by multiplying one frequency by three and subtracting twice the other frequency. Another method is to sample the transmitter output with a selective receiver and an accurate S meter. Tune the spectrum near the desired signal for responses at the third and fifth order frequencies. The ratio of these responses is then compared to the carrier frequency, and an S/D power ratio is derived. Twenty-five DB is the minimum S/D ratio.

BASIC ELECTRICITY

34. Show the formulas for resistors, capacitors, and inductors in series and in parallel.

Resistors in parallel: $\dfrac{R1, R2}{R1 + R2}$

or for more than 2,

$$\dfrac{1}{\dfrac{1}{R1} + \dfrac{1}{R2} + \dfrac{1}{R3}}$$

Resistors in series, add R1, R2, R3, etc.

Capacitors in series: $\dfrac{C1, C2}{C1 + C2}$

or for more than 2,

$$\dfrac{1}{\dfrac{1}{C1} + \dfrac{1}{C2} + \dfrac{1}{C3}}$$

Total inductance in series add L1, L2, L3, etc.

Inductance of 2 inductors in parallel: $\dfrac{L1, L2}{L1 + L2}$

or for more than 2,

$$\dfrac{1}{\dfrac{1}{L1} + \dfrac{1}{L2} + \dfrac{1}{L3}}$$

35. What is the total opposition to the flow of AC which includes capacitive reactance, inductive reactance, and resistance?
    Impedance.
36. What is the opposition to AC current offered by a capacitor, by an inductance?
    Inductive and capacitive reactance.
37. Which reactance is inversely proportional to frequency?
    Capacitive reactance.
38. What determines the voltage drop across series connected resistors, capacitors, inductors?
    The same current flows in a series circuit so voltage drop is proportional to impedance and current. In resistors, the voltage drop is equal to the current multiplied by the resistance. The proportion of the applied voltage across each resistor equals the individual resistance divided by the total resistance.
    When using inductance, the inductive reactance is proportional to the inductance. Inductance may be used instead of reactance. However, capacitive reactance is inversely proportional to capacitance, so the individual voltage drop across a capacitor in series is the inverse of the ratio of one capacitor to the total of all the capacitors.
39. The ratio of current and resistance in any circuit is also known as what?
    Ohms Law.
40. R E/I and E/IR is also known as what?
    Ohms Law.
41. In AC circuits, how would ohms be expressed?
    R, or ohms, becomes Z, or impedance.
42. What does the following formula express?
    DB 10 log P2/P1
    Decibel or power ratio. P1 and P2 are the two power levels being compared. A power ratio of 10 is 10 DB, of 100 is 20 DB, and of 1000 is 3 ODB, and 3 DB is a

doubling of power. Power gain or loss can be added or subtracted.

43. What is an "S" unit?

    It is a reference composed of an arbitrary number of decibels. Receivers have an S meter with 0 to 9 divisions. Above S9, the meter is calibrated in decibels. Each S unit is from 5 to 6 decibels.

44. If you adjusted the impedance of a load to its power source so they are equal, what would you be doing?

    Impedance matching.

45. Why is it necessary to match impedances?

    To allow the greatest power transfer. The antenna impedance must be matched to its transmission line, and its transmission line impedance matched to the transmitter. Power loss and harmonic radiation on undesired frequencies result when impedances are not correctly matched.

46. Name the following component. A coil has a current going through it, a changing magnetic field surrounds the turns of the coil. Another coil, close to the first, receives mutual induction. The first coil is the primary, the second, a secondary.

    Transformer.

47. Give the formula to determine a transformer's output voltage, when its input voltage and the turns ratio are known.

$$E_s \text{ (secondary voltage)} = \frac{N_s \text{ (secondary turns)}}{N_P \text{ (primary turns)}} E_p \text{ (input voltage)}$$

48. Give the formula to find the impedance looking into the primary from the source of power. Assume the load impedance and the turns ratio are known.

$$Z_p \text{ (primary impedance from power source)} = Z_s \begin{pmatrix} \text{impedance} \\ \text{of load con-} \\ \text{nected to} \\ \text{secondary} \end{pmatrix} \begin{pmatrix} \dfrac{N_p \text{ primary turns}}{N_s \text{ secondary turns}} \end{pmatrix}$$

49. Name the type of resonant circuit described here. The impedance equals the vector sum of resistance, inductive, and capacitive reactance. At resonance, capacitive and inductive reactances are equal and opposite and cancel, leaving the resistance as the effective impedance. This resistance is very low.

    Series resonant circuit.

50. Describe the parallel resonant circuit.

    Its impedance equals the square of the reactance of either capacitive or inductance reactance. The parallel impedance is resistive and is a very high value.

51. Give the formula for finding the resonant frequency.

    frequency in Hertz =
    $$\frac{1}{2\pi \sqrt{L \text{ (inductance in Henrys)} \times C \text{ (capacitance in farads)}}}$$

52. What is Q? Give the formula for finding Q.

    The Q of a circuit is its quality factor. The selectivity, or how sharply a circuit tunes, depends on the Q of the reactance in this circuit. In series resonant circuits, Q equals reactance divided by resistance (X/R). X can be the reactance of either L or C. R is the total series resistance. In a loaded parallel resonant circuit, Q = R/X.

53. What are the advantages of resonant circuits in amateur transmitters?

    In RF amplifiers, resonant circuits reduce the cathode-plate capacitance and give the high selectivity necessary for harmonic reduction. RF amplifier efficiency and gain is sharply increased by resonant circuits. Thus reduced driving power is required and greater power output can then be obtained.

54. What type of circuit is described here? Some of the output energy is sent back in phase to the input.

    An oscillator is described. Where undesirable oscillation takes place, neutralization is required.

Neutralization occurs by feeding some of the tube's energy back to the input OUT of phase. Because of this phase relationship, the cathode-plate capacitance is cancelled out.

55. List the distinguishing characteristics of class A, B, and C amplifiers, their applications, and the efficiencies obtained by each.

    Class A plate current flows throughout the entire cycle. It is used where low distortion is necessary such as audio amplifiers. Its efficiency is only about 20 percent.

    Class B plate current flows during the positive half of the signal-input cycle. It will produce more power than the class A, but with increased distortion. It is used in push-pull audio power amplifiers and as a linear RF power amplifier.

    Class C plate current flows in pulses that are less than one-half of the signal input cycle. It has the highest efficiency—up to 80 percent. It is used essentially for radio frequency amplifiers where some distortion is permissible.

56. What precautions would you take when constructing an antenna to minimize harmonics?

    Poor electrical connections act as rectifiers and produce harmonics. Protect all connections against the weather. Use a transmatch or other impedance matching device to discriminate against undesirable harmonic frequencies.

57. How would you design an RF amplifier to minimize harmonics?

    Use shielded leads for DC and filaments. Install RF filters in all leads leaving the cabinet and shield the entire unit. Use high Q tuned circuits, neutralization, and a low-pass filter in the antenna and transmission line.

58. What undesirable possibility is being discussed here? It can be prevented by feeding some of the output back to the input OUT of phase, by neutralization.

Oscillation in an RF amplifier.

59. How would you proceed to neutralize an RF amplifier?

    Remove the plate and screen grid voltage. Input and output resonant circuits are tuned to the oscillator or buffer frequency. This is done by reading a sensitive RF indicator which is coupled (with a wire link) to the amplifier tank circuit. The neutralizing capacitor is adjusted until absolutely no RF power exists in the plate circuit.

60. Explain the difference between these 3 types of amplifiers: grounded cathode, grounded plate, and grounded grid.

    Grounded cathode amplifier has high input impedance, high power gain, but care must be taken to prevent unwanted oscillation. Grounded plate (cathode follower) has high input impedance and low output impedance. It is used to match high impedance circuits to low impedance circuits. The grounded plate and grounded grid amplifiers have their output in phase with their input. However, the output and the input of the grounded cathode amplifier are 180 degrees out of phase. The grounded grid amplifier has low input impedance, moderate power gain, and need not be neutralized.

61. A dipole, or doublet, or half-wave antenna is 2 to 10 percent shorter than a half wavelength in space. Give the formula for calculating a half-wave and a quarter-wave antenna.

    $$\tfrac{1}{2} \text{ wavelength} = \frac{468}{\text{freq. (MHz)}} \qquad \tfrac{1}{4} \text{ wavelength} = \frac{234}{\text{freq. (MHz)}}$$

62. List the elements of both a transistor and a triode vacuum tube. Compare them to each other.

    Transistor—base, emitter, collector.
    Tube—filament, cathode, grid, plate.

    The transistor has no equivalent to the filament. The emitter is the equivalent of the cathode, the base compares to the grid, the collector compares to the plate.

63. What device is being discussed? A small voltage is impressed on the base. The device regulates electron flow from emitter to collector.

    The transistor.

64. What device is being discussed? Current flows only in one direction, and a high resistance exists in the opposite direction. The anode and cathode of this device compare to the tube's plate and cathode.

    Diode.

65. If we wished to design a transformer which was very compact, was self-shielding, and had a high Q, what core material would be used?

    We would use a doughnut-shaped ferrite or iron toroid core.

66. What device is being discussed? It has aluminum foil plates with a thick liquid separation. The insulating dielectric is a thin film of insulating material forming on one set of plates by electrochemical action. The capacitance can be very large compared to capacitors using other dielectrics. This device is used in power supply filters due to its high volume to capacitance ratio.

    The electrolytic capacitor.

67. How does the quarter-wave ground plane antenna differ from the horizontal half-wave dipole?

    The radiation from the quarter-wave ground plane is omnidirectional, has a low angle of radiation, and is vertically polarized. Ground plane radials are required. The horizontal dipole does not require ground radials, since it uses ordinary ground as a reflector. The dipole radiates a horizontally polarized, bidirectional signal. If the dipole is close to ground, its angle of radiation is high, its directivity and impedance both decrease under these conditions.

68. What determines the characteristic impedance of a transmission line?

The diameter of the conductors, the spacing between them, and to a lesser extent, the insulating material between them. To find the characteristic or surge impedance use the formula:

$$Z_o \text{ (characteristic impedance)} = 276 \log \frac{b \text{ or distance between conductors}}{a \text{ or radius of the conductor}}$$

69. What is the effect of increasing the line spacing or decreasing the wire conductor size on a transmission line's impedance?

The impedance is increased. When air is not the dielectric, the dielectric must be multiplied by $1/\sqrt{K}$, where K is the relative dielectric constant of the material used. For plastics used in transmission lines, the value of K is 2.2.

70. What is SWR? How can SWR on a transmission line be determined from the incident and reflected voltages?

When transmission lines are not perfectly matched, the variations of current and voltage along the line (SWR) is the ratio of maximum to minimum current along the line. To measure SWR, a bridge circuit or reflectometer measures both forward and reflected voltage. Use the formula:

$$\text{SWR} = \frac{V_f \text{ (forward voltage)} + V_r \text{ (reflected voltage)}}{V_f \text{ (forward voltage)} - V_r \text{ (reflected voltage)}}$$

71. Is it possible to use a center-fed non-resonant antenna on different amateur bands?

Yes, the open wire feedline can be used with a transmatch (impedance matching tuner) connected between the transmitter and the feedline. The transmatch will tune the entire system to resonance at the operating frequency, balance an unbalanced line, and reject harmonics in the transmitter output.

72. Give and explain the formula for determining the characteristic impedance of an air dielectric, coaxial transmission line.

Z_o (characteristic impedance) =

$$138 \log \frac{b \text{ or inside diam. of outer cond.}}{a \text{ or outside diam. of inner cond.}}$$

When a solid dielectric material (other than air) is used as an insulator, the velocity factor must be calculated. The result in the formula above should be multiplied by $1/\sqrt{K}$, where K is the dielectric constant (usually 2.2) of the dielectric material.

73. In an SSB transmitter, how is the input power to the plate circuit of the final amplifying stage determined?

The product of DC plate current and DC plate voltage during the largest voice peak (shown on the plate current meter) gives the DC input power. It may not exceed one kilowatt, which includes *all* stages supplying power to the antenna. For example, if the last RF stage is grounded-grid, the driver stage contributes RF power to the antenna, so the power of the driver stage must be added to determine the total power input.

74. Show how the oscilloscope is used to adjust a radio telephone transmitter.

The RF output of the transmitter is displayed on the oscilloscope screen. The modulation envelopes shown below determine how heavily to drive the RF amplifier.

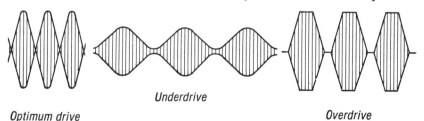

Optimum drive   *Underdrive*   *Overdrive*

Figure 13-1

The amount of linearity is seen by comparing these patterns to a pure half cycle of a pure sine wave. With a phasing transmitter, a "bow-tie" or double triangle pattern can be displayed. In the "bow-tie" test, curvature of the sloping sides of the triangle shows distortion. Adjust the transmitter for straight sides. In the illustration, the optimum pattern does not yet have flattened tips as when overdrive occurs. With underdrive, the RF amplitude is low.

To check the audio-frequency amplifier, the input terminals of the oscilloscope are connected in parallel with the input terminals of the amplifier being tested, and are connected to the horizontal amplifier input terminals in parallel with the output of the amplifier being tested. Adjust the vertical and horizontal amplifier gain so that both signals show the same deflection. A linear audio amplifier will show up when the voice pattern is a sloping straight line of varying length, provided the input and output phase relationship is either zero or 180 degrees. If there is no distortion, a series of concentric smooth ellipses, varying in size with voice amplitude, occurs with intermediate phase relationships. A bend at some point in the line or irregularity in the ellipses indicates nonlinearity. To check the linearity of the oscilloscope itself, connect its input terminals in parallel and apply a single tone. For SSB where the linearity of the transmitter RF amplifier is critical, sample its grid (input) and plate (output) voltages with link coupling to those circuits. When using inexpensive oscilloscopes, RF signals must be applied directly to the deflection plates, since the horizontal and vertical amplifiers will not work at high frequencies. To overcome this, the two RF voltages may be rectified to produce two audio signals for comparison. This procedure has already been described.

75. To eliminate the possibility of electric shock, what precautions should be taken in the construction and operation of amateur equipment?

Always use an external ground (water pipe), never use AC/DC transformerless power supplies. A metal enclosure will prevent contact with the electrical circuit. Use three conductor power cables having a ground lead, and use adequately rated components and wiring. This prevents overheating of wires and insulation breakdown causing high voltage to appear at unexpected points. Always use bleeder resistors across large capacitors in power supplies to discharge any voltage remaining when the equipment is turned off. Always use an interlock switch which cuts off power to circuits when equipment is opened.

76. Draw all stages of an SSB transmitter of the filter type. Explain the function of each stage.

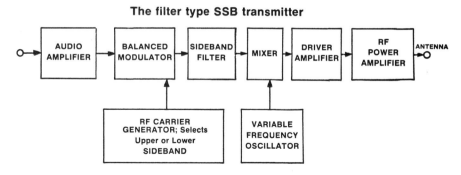

Figure 13-2

In this circuit, both RF oscillator signals and audio are fed into a balanced modulator. Its purpose is to suppress the RF carrier. Both sidebands are then applied to a band-pass filter. Its high selectivity permits one sideband to pass while the other is rejected. A sideband signal generated at a low frequency can be converted to the desired transmitting frequency by a heterodyne oscillator and mixer with its output tuned to the desired channel.

77. Draw a block diagram of a transmitter for F3 emission. Explain the function of each stage.

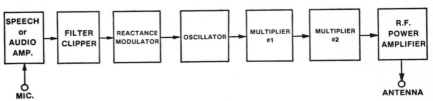

Figure 13-3

The clipper-filter clips audio peaks slightly to increase the average audio power, and also filters out high audio frequencies which produce noise. The reactance modulator shifts the oscillator frequency according to the audio signal. The oscillator provides an RF signal at a subharmonic (much lower frequency) of the output frequency. Frequency multipliers bring the oscillator frequency up to the output or operating frequency by generating harmonics. The Class C PA (power amplifier) increases RF power to the desired level.

14

Repeaters and How to Understand Them

FINDING REPEATERS

When you have passed your Technician's Class exam and received your Technician's license, you are ready to make voice transmissions on the VHF bands. One of the most fascinating aspects of these bands is the repeater.

Even before you have taken any FCC exam, you may eavesdrop on any inexpensive radio with a Public Service band. You will notice between 144 and 146 MHz that there are one or more very strong signals that appear to be two amateurs talking to each other. After listening to this frequency for some time, you will find many hams using it. It will not be long before you will notice something strange. One of these amateurs will be talking over a walkie-talkie 80 miles south of where you are. He will be talking to another ham 50 miles north of your location, and she will be transmitting from an auto. How can a walkie-talkie (handie-talkie) have a 130 mile range?

THE MACHINE

This miracle of modern amateur radio is made possible by the repeater, nicknamed the machine. The machine is located on the top of a mountain or on the top floor of a skyscraper. A local amateur radio club (or repeater club) has raised the money, found a high location, and installed the machine. The machine has a very sensitive FM receiver. This receiver has its audio output directly connected to the audio input of a transmitter. The transmitter operates on a second frequency close to the first one. The result is that everything received on the first frequency is retransmitted on the second frequency.

The repeater has a carrier-operated relay—COR. This device is connected to the receiver and turns the transmitter on only when a signal is being received by the sensitive receiver. So when you are holding a one watt walkie-talkie in a poor location, you can trigger a 500 watt transmitter on the top of a mountain or atop the Chrysler building in New York City (WR2ACD).

The ARRL supplies to its members a free repeater directory listing the receiving and transmitting frequency of every repeater in the country. Their 1977 repeater directory listed some 4,000 repeaters. That was a 30 percent increase over the year before. So you can see how popular repeaters are becoming all over the country.

The majority of repeaters operate on 2 meters (144-146 MHz). You may find an isolated one here and there on 6 meters (50-52 MHz). Repeaters on 1¼ meters (220 MHz) are becoming more common each year. Several repeaters are now operating up on 70 centimeters (432 MHz).

It is not difficult to see why repeaters are becoming so popular. The price of handie-talkies has been dropping each year. One manufacturer sells a handie-talkie kit for $125. You can be inside a steel building or on a city street and talk to someone 100 miles away. Using a touch-tone pad on the handie-talkie, you could trigger a repeater and then dial any telephone number anywhere in the world. Of course, you would have to be a member of the repeater club to do this. Suppose you made $100 phone calls to Australia every day. These calls would appear on the phone bill of the repeater club. They would not survive too long having to pay for all the long-distance calls you made over their repeater. As a member of the repeater club, you would pay for these phone calls yourself. Under these conditions, you will have second thoughts about dialing Australia every day.

REPEATERS AND HOW TO UNDERSTAND THEM

About 90 percent of the repeaters in the U.S. are open machines. This means any amateur broadcasting on the repeater's input frequency (the frequency its receiver is tuned to) can trigger the repeater. Any transient mobile operator driving in the repeater's range can trigger the machine. Some machines even have a beep device. Before you key the machine you wait for the beep, and the time you have waited allows another station to break in. Approximately 10 percent of the repeaters in the U.S. are closed machines. This means only members of that particular repeater club can trigger that machine. Their transmitters have a special secret tone which

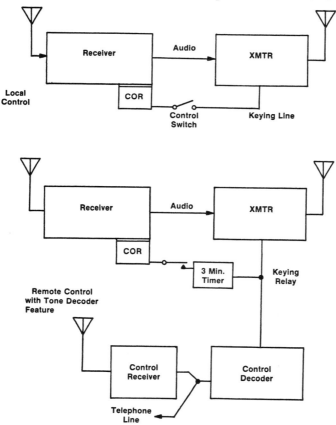

Figure 14-1 *In local control position, the repeater can be serviced. However, the normal condition is the remote position.*

enables them to open up the repeater. Closed repeaters are on the decline and are discouraged by most thoughtful amateur groups.

A block diagram in Figure 14-1 shows a repeater set up for both local control and remote control. The normal situation on the air is remote control. However, on weekends, members of a repeater club will often spend some time at the repeater location tuning up the rig and performing preventive maintenance.

Repeaters are basically duplex machines. That is, simultaneous transmissions take place between two stations using two frequencies. Two amateurs may wish to leave the repeater frequencies for a prolonged conversation by themselves. They would then select a frequency for simplex operation. On this frequency, alternating transmissions between these two stations could take place for a long period of time if so desired, and the repeater would be left free for the use of many other stations.

With the ability of a handie-talkie to trigger a high power repeater station atop a mountain, it goes without saying that responsible operating practice becomes a necessity.

The ARRL has outlined 14 rules for responsible repeater operating procedure. (See Table 14-1 next page.)

REMOTE BASE STATIONS

A repeater has two separate frequencies, an input and an output frequency. They are usually installed by a repeater radio club. Another form of operation is possible, which has some similarity to the repeater principle. This is remote base operation. Let's say your location is very poor and you would like to operate on the VHF frequencies. You may install a remote base station at an ideal location. Unlike a repeater, this station would be for your exclusive use. Figure 14-2 shows the block diagram of such a station. It is simplex. That is, both the transmitter and receiver operate on the same frequency. This is possible because they are never turned on at the same time. At the control station, you decide which should be turned on—the transmitter or the receiver.

REPEATER TIPS

DO'S

DO KEEP ALL TRANSMISSIONS SHORT. Emergencies don't wait for monologues to be finished. If you talk to hear your own voice, what you want is a tape recorder, not an FM rig.

DO THINK BEFORE YOU TRANSMIT. If you can't think of anything worth saying, don't say anything.

DO PAUSE A COUPLE OF SECONDS BETWEEN EXCHANGES. Someone with a high-priority need for the repeater may want to break in; also, some repeaters are configured so that a brief pause in transmissions is necessary to reset the three-minute timer.

DO IDENTIFY PROPERLY. "W6XYZ mobile" is not enough, even if you're 300 miles from another call area; "W6XYZ mobile 6" is required. You must give the call of the station with whom you were in contact at the end of the contact.

DO BE COURTEOUS. A repeater is like a telephone party line, and requires the same kind of cooperation in its use.

DO USE SIMPLEX WHENEVER POSSIBLE. Leave the repeaters available for those who need them.

DO USE THE MINIMUM POWER NECESSARY to maintain communication. Not only is this an FCC requirement, it's also common courtesy.

DO SUPPORT YOUR LOCAL REPEATER CLUB, even if it doesn't require all users to be members. Maintaining a good machine is an expensive undertaking, and you should do your share.

DON'TS

DON'T ABUSE AUTOPATCH PRIVILEGES. Business messages are not permitted in the amateur service. Don't force the control operator to terminate your call in order to avoid a violation.

DON'T BREAK INTO A CONTACT unless you have something to add. Interrupting is no more polite on the air than it is in person.

DON'T FORGET THAT AMATEUR RADIO IS ALLOCATED FREQUENCIES BECAUSE IT IS A SERVICE, not just a hobby. Don't neglect the public service aspects of VHF, FM communication, such as accident reporting, emergency preparedness, etc.

DON'T TRY TO PROVE WHAT A GREAT OPERATOR YOU ARE by criticizing the operating techniques of others on the air. Instead, set an example which others will be proud to follow.

DON'T MONOPOLIZE A REPEATER. The best repeater users are the ones who do a lot of listening, and little transmitting.

DON'T FORGET THAT WHAT YOU SAY OVER A REPEATER CAN BE HEARD OVER THOUSANDS OF SQUARE MILES by anyone with an inexpensive "public service band" monitor. These people are potential hams; if they like what they hear on the air, they will want to get licenses and join us. Don't leave them with a bad impression of our hobby by making thoughtless or off-color remarks.

Table 14-1

244 REPEATERS AND HOW TO UNDERSTAND THEM

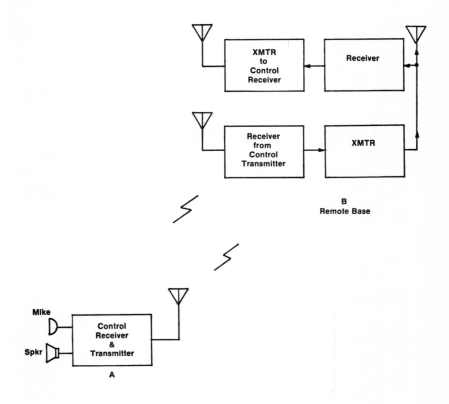

Figure 14-2 *Control station (A) is at a poor location. The remote base (B) is located at an excellent location. The control receiver is receiving from the remote xmtr. The control transmitter sends to the remote receiver.*

15

Key Factors in Amateur Test Equipment

INTRODUCTION

If you construct your own amateur gear, you must be able to repair it yourself. Even if you purchase all your equipment, it is still a good idea to be able to repair it yourself. When loading up an antenna, an SWR meter will come in handy. When you receive a report that your voice signal is distorted or your telegraphy signal has a chirp, it is time to trouble-shoot your equipment. Whenever you make any changes in your antenna, a field strength meter must be used.

Inexpensive oscilloscopes usually cannot measure high frequency waves. An RF probe will permit your low cost oscilloscope to do some of the work of an expensive oscilloscope. If you have a low price receiver, a frequency marker is a necessity. Otherwise, if you have to keep an appointment at a specific frequency, the unreliable calibration of your receiver will make this impossible. When you are aligning your receiver for maximum sensitivity, a noise generator will come in handy.

ELECTRON VOLTMETERS

Whenever a voltage is measured, always find out first what the resistance is of the circuit being measured. The input resistance of the voltmeter you use must be at least ten times that of the parallel resistance of the circuit being measured. If this rule is not followed, the reading obtained will be below the actual voltage because the voltmeter will load down the circuit. For example, if a DC voltmeter has a resistance of 20,000 ohms per volt, and its full-scale range is 10 volts, then its DC resistance will be 200,000 ohms (20,000 ohms for each volt on the meter scale). In this instance, this meter could not be used in a circuit whose resistances were over 20,000 ohms.

Small, inexpensive volt-ohm meters may not be able to measure small AC currents (milliamps) and small voltages (millivolts). They may also burn out easily when the wrong meter scale is used.

To overcome the problem of having an extremely high resistance input, VOMs are now made with electronic circuits. For example, the FET-VOM will typically have an input resistance of ten million ohms. It can thus be used in circuits having resistances up to one million ohms.

RF PROBE

The FET-VOM or any other high resistance input voltmeter is satisfactory for measuring DC voltages. Instrument rectifiers are also included in such instruments to facilitate the measurement of AC voltages. However, when an attempt is made to measure high frequency (RF) voltages, the typical FET-VOM is not satisfactory. RF in the lead connected to the instrument will have enough capacitance to ground to short circuit some of the RF being measured. The inductance in the wire will also resist the RF being measured. To eliminate these problems, an RF probe can be built or purchased. Figure 15-1 shows a small RF probe built into a metal cigar tube.

The metal tube is at ground potential so your hand does not affect the voltage reading. The series resistor in the probe prevents RF from being fed through the probe cable, and consequently being dissipated. This value (4.3 m) and the input resistance of the FET-VOM (10 megohms) form a voltage divider. This reduces the peak rectified value to the RMS or true value of the voltage. The meter reading is now consistent with other AC readings displayed on the same meter.

AMATEUR TEST EQUIPMENT

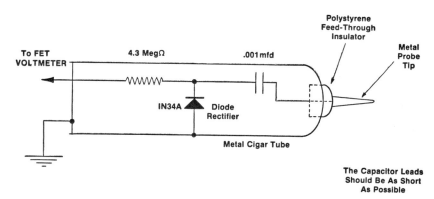

Figure 15-1 *Voltages up to the VHF region can be accurately measured with this RF probe. The rectifier is a point-contact germanium with no more than 1 mmfd capacitance. A pin diode may be used.*

This voltage probe can also be used to measure the actual power of a radio transmitter. The probe is connected across the load which is absorbing the transmitted output. It might be connected across the antenna terminals if one terminal is grounded. If the load is a pure resistance or the transmission line is a pure resistance, then the following formula is used to find the transmitter's power: E^2/R. If the transmission line is 72 ohms and the voltage is 8.5 volts, then:

$$8.5 \times 8.5 = 72 \text{ ohms}/72 = 1 \text{ watt}$$

A higher voltage would indicate higher power.

This measurement indicates only apparent power, when the line is not a pure resistance. Another way of measuring power is to use a 40 watt light bulb across the antenna terminals of a 60 watt transmitter. When the bulb is fully lit (40 watts = 70 percent of 60 watts), then the transmitter is producing full power at reasonably high efficiency.

THE OSCILLOSCOPE

We have seen how the VOM can measure both DC and AC voltages. By adding an RF probe, voltages into the very high frequencies may also be measured. When looking at an AC or RF voltage, we only see the averaged potential over a long period of time—the relatively slow time constant of the meter.

Unfortunately, we have been missing a great deal of information. We have not seen what the AC or RF waveshape looks like. In other words, how the voltage changes with respect to time. Earlier we learned about the sine wave. When looking at an AC voltmeter, the RMS value, or 0.707 percent of peak value, is what our meter reading shows.

Using an oscilloscope, we can see sine waves, distorted waves, transient waves, linear waves, and lissajous patterns. The heart of the oscilloscope is the cathode-ray tube. Similar to the picture tube in your TV set, a tiny dot is projected onto a phosphor-coated screen. The CRT in your TV set uses vertical and horizontal coils (yokes) to deflect the dot up and down, and from left to right. The oscilloscope CRT applies a voltage to electrostatic deflection plates. These plates then deflect the dot up or down, or from left to right.

Figure 15-2 is a block diagram of an oscilloscope showing the vertical and horizontal deflection plates of the CRT. The two vertical plates on the left and right move the dot horizontally, while the two horizontal plates (top and bottom) move the dot up and down. The sawtooth generator generates a slowly ascending linear wave which drops sharply, then resumes its linear motion. This sawtooth wave shown in Figure

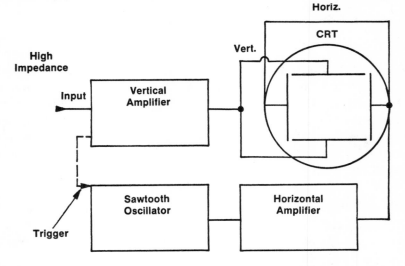

Figure 15-2 *Block diagram of the basic oscilloscope. The cathode-ray tube is enclosed in the circle.*

15-3 causes the dot to move from left to right. This is the time base. If the sawtooth takes 200 milliseconds from its start at A to its completion at B, then we have a 200 millisecond time base. This means we can see any voltage variation occurring over this 200 millisecond period. The input waveform we wish to observe is sent to the vertical amplifier, and then pulls the dot up and down. As the dot moves very fast, it appears to the eye as a line. This line fluctuates over a predetermined time base and becomes the waveshape we wish to observe.

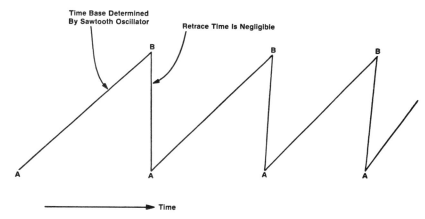

Figure 15-3 *The saw-tooth generated in the oscilloscope. The time elapsed from A to B is known as the time base.*

In Figure 5-2, the sawtooth circuit generates the horizontal baseline. If we disconnect the sawtooth generator, we can apply a second input signal to the vertical plates. The difference in amplitudes, frequency, and phases of the two voltages then appear as a lissajous pattern. The shape of the pattern indicates the frequency ratio between the two inputs. In Figure 15-4 we see various lissajous patterns. These patterns are useful to calibrate audio frequency generators to be able to generate a given tone. More and more amateur equipment uses audio tones to open repeaters, to turn on remote equipment, etc. The unknown audio tone is fed to one set of deflection plates, and the 60 Hz power line frequency is fed to the remaining set of plates. Their respective ratios can then be observed.

One of the better known oscilloscope testing procedures is determining the modulation percentage of an amplitude

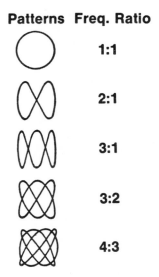

Figure 15-4 *Lissajous figures and corresponding frequency ratios for a 90° phase relationship between the voltages applied to the two sets of deflecting plates.*

modulated (AM) transmitter. In this test, the various patterns produced are shown in Figure 15-5. Each corresponds to a given level of modulation. When applying RF energy to an oscilloscope, always make sure the vertical and horizontal amplifiers can pass the RF frequency. Many inexpensive oscilloscopes will pass up to 5 MHz. This means that if 7 MHz (40 meters) must be applied to the oscilloscope, it must be applied directly to the deflection plates. A worthwhile oscilloscope should have a trigger input terminal. When looking at high frequencies, apply some of the RF to the trigger input terminal. This will control the starting time of the sawtooth time base and the picture will be clear and sharp.

FREQUENCY MARKER

Unless you have an expensive receiver, the calibration of its tuning dial cannot be completely trusted. The low end of the 80 meter band is 3.5 MHz. If you broadcast exactly on this frequency you are too close to the band's edge, and the FCC

AMATEUR TEST EQUIPMENT

251

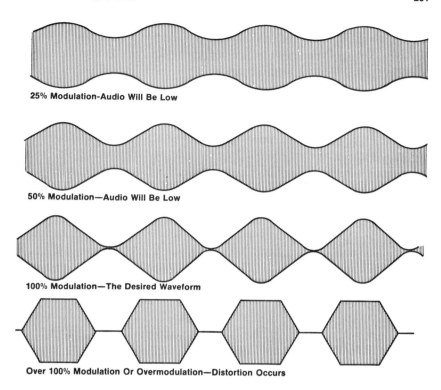

Figure 15-5 *The various oscilloscope displays indicate the level of modulation of an amplitude-modulated transmitter.*

may send you a citation of a violation. It is always best to know exactly where band edges are to avoid unpleasant situations.

Morse code training and information nets meet on a given frequency. For example, every Tuesday night at 9:00 PM, the AMSAT net broadcasts on 3850 kHz. AMSAT is the acronym for Amateur Satellite Corporation. Operators on this net will let you know whether satellite launches were successful, and what is the current status of the various amateur satellites. Under crowded band conditions, it is often difficult to find a given frequency if your receiver's frequency calibration is even slightly off. A good frequency marker will overcome this problem.

A frequency marker is simply an oscillator designed to be very rich in harmonics. The oscillator's fundamental frequency is 100 kHz, and it radiates a signal every 100 kHz thereafter up to 10 MHz. This oscillator's output is connected directly to the receiver's antenna terminal. Figure 15-6 shows the circuit of the 100 kHz frequency marker. The FET oscillator can be powered with a tiny 9 volt radio battery. Using the receiver's beat frequency oscillator (BFO), zero beat the oscillator against WWV on 5 MHz or 10 MHz. This is done by critical tuning of capacitor C1. When you tune your receiver through the amateur bands (with the BFO left on), a heterodyne whistle will be heard every 100 kHz. When the whistle decreases in pitch until the sound is a hoarse wobulating, then you are exactly zero beat. The oscillator and your receiver are tuned to precisely the same frequency. Knowing this, the frequencies between the 100 kHz divisions of your receiver's dial can then be determined. For example, when tuning 80 meters, a whistle will be heard as you approach 3.5, 3.6, 3.7, 3.8, 3.9, and 4 MHz.

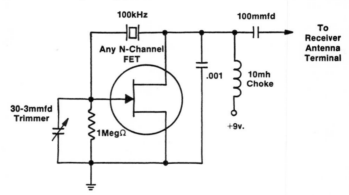

Figure 15-6 *Simple 100 kHz oscillator circuit produces a marker signal every 100 kHz apart for receiver calibration.*

AUDIO OSCILLATORS

A simple audio-frequency oscillator is useful to generate the stable audio tone for checking the audio quality of a transmitter. Two such oscillators at different frequencies are used for the two-tone test for testing single sideband transmitters. Audio oscillators also make good signal tracers for audio amplifiers. Simply connect the oscillator to the input of the last stage. If you hear its tone in the speaker, then reconnect

AMATEUR TEST EQUIPMENT

the oscillator further back in the circuit until you reach the point where the tone disappears. The defective component is between the point where the tone was last heard and the point where it disappeared. Figure 15-7 illustrates the most simple type of audio oscillator. No transformer is required and the entire unit can be built inside a metal cigar tube. A small 9 volt radio battery taped to the cigar box can provide power to the oscillator. The cigar tube becomes an audio probe.

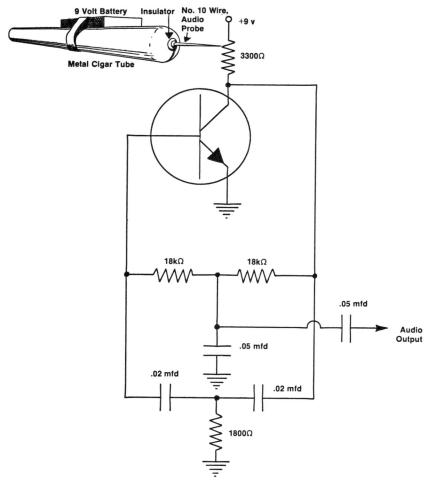

Figure 15-7 *The twin-T oscillator does not require a transformer. It will find many applications in audio testing.*

FIELD STRENGTH METER

The field strength meter in Figure 15-8 is simply a crystal broad-band receiver using a sensitive meter to indicate output level. This handy device will find 100 uses around your radio shack. Placed near a transmitter circuit, the transmitter can be tuned for maximum output by obtaining a maximum reading on the field strength meter. Placed near an antenna, the antenna or the transmitter can be adjusted for a maximum reading of the field strength meter. The author once placed one of these gadgets on his auto dashboard. Every time the push-to-talk button of my transmitter microphone was pressed, the field strength meter would show it. As it actually receives its signal through the air, there is never any doubt that your transmitter is working properly. The longer its antenna, the more sensitive it is. The antenna may be made out of number 10 electrician's wire or out of 1/8 inch welding rod. It is important that the antenna be rigid, otherwise changes in meter readings will occur.

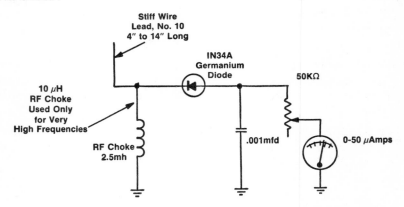

Figure 15-8 *The field strength meter is very valuable for tuning up transmitters and antennas.*

When you first install a new transmitting antenna, it is a good idea to put the field strength meter someplace near the antenna where a given reading is obtained. Make a note of the meter reading and the precise spot where the reading was obtained. After a year, repeat the procedure to determine if the

AMATEUR TEST EQUIPMENT

255

same reading is again obtained at the same place. If the reading is lower, then the elements and the weather have created dirty contacts, or the antenna insulators need to be cleaned. In either case, your transmitted signal is not getting out as well as it was when the antenna was new. Yet the meters in your transmitter circuits will show no indication of this.

MEASURING VSWR

Remember that problem we often encountered when trying to feed RF energy into an antenna? If the transmission line impedance was not properly matched to the antenna impedance, the antenna would not accept the RF energy. Some, or all of it, would bounce back through the transmission line as reflected power. The ratio of forward power to reflected power, then, became an important characteristic of our antenna system. The Voltage Standing Wave Ratio had to be maintained as close as possible one to one, the ideal VSWR.

A VSWR indicator shows the ratio of forward to reflected energy in an antenna transmission line. The VSWR circuit in Figure 15-9 is installed between the coaxial transmission line and the transmitter. An insignificant amount of energy is coupled (through T1) from the coaxial line into the VSWR indicator. With the switch (S1) in its forward position, forward power is indicated on the 50 micro-ampere meter. Forward power is the position that gives the highest reading. This SWR calibration procedure should be done using a 50 ohm power resistor as a dummy load. The dummy load is the equivalent to the perfectly matched antenna, so we know that the VSWR will be 1:1, or the ideal. Make sure that the dummy load (power resistor) has a power rating which will handle all the transmitter power. If the transmitter is 25 watts (70 percent efficiency), then a 20 watt resistor is satisfactory.

A dummy load is necessary when tuning up a transmitter. Tuning up a transmitter into an antenna will cause interference to other stations and should be avoided. With the dummy load connected, switch S1 into the "reflected" position. Now, adjust C1, the variable capacitor nearest the transmitter connection, until the meter reading drops to zero. This is called a null. Then reverse the transmitter and load connections and repeat the

C1, C2 = 0.5-5 pF trimmer
CR1, CR2 = IN34A or equivalent
MI = 50 µA meter
R1 = ½ watt, 25,000 Ω
R2, R3 = 33Ω, ½ watt
RFC = 1 mh RF choke
S1 = spdt toggle switch

T1 = 60 turns No. 28 enamel wire close wound on Amidon T-68-2 toroid core (secondary). Primary is 2 turns of small diameter hookup wire over T1 secondary.

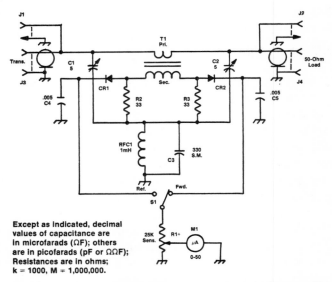

Except as indicated, decimal values of capacitance are in microfarads (ΩF); others are in picofarads (pF or ΩΩF); Resistances are in ohms; k = 1000, M = 1,000,000.

Figure 15-9 *This SWR meter indicates how much transmitter power is going into the antenna and how much is reflected back through the transmission line.*

meter null adjustment. This time adjust C2 for the meter null. Once you have tuned both C1 and C2 for a null in both directions, the transmitter connection and the load connection may be used.

When an antenna is connected to the 50 ohm load connection, set the meter sensitivity potentiometer for a full-scale reading with S1 in the forward position. If the VSWR is close to 1:1 (almost no reflected power), then the meter will read almost zero with S1 in the reflected position. If the reflected power shows a noticeable reading, then the antenna is not tuned properly or the transmission line is not terminated in its characteristic impedance—50 ohms. An old trick is to test VSWR at both the high and low end of an amateur band. If the VSWR looks better at the high frequency end of the band, then the antenna is too short. If the VSWR looks better at the low frequency end of the amatuer band, then the antenna is too long.

NOISE GENERATORS

At the high frequencies, it is often difficult to tune a receiver for the greatest sensitivity. This is so because noise is generated in the RF amplifier, mixer, and oscillator stages of super-heterodyne receivers. This noise sometimes makes it difficult to hear signals coming in from the antenna. A diode noise generator takes advantage of the fact that electrons flowing through a crystal diode create a shot effect, producing a hiss type of noise in the RF spectrum. In Figure 15-10, a small 9 volt radio battery is the DC source that sends electrons flowing through the IN21 or IN23 crystal diode. R2 provides the current return path to the battery. R1 is adjusted to increase or decrease the amount of noise. When the noise generator's output is connected to the antenna terminal of a receiver, a loud hissing noise should be heard. This noise is external to the receiver's circuits and would be the equivalent of a constant noise coming from the antenna. Thus we tune the RF, mixer, and IF stages in the receiver for the loudest noise from the generator.

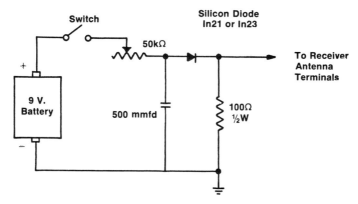

Figure 15-10 *This diode noise generator has several interesting applications.*

There is no danger that we are tuning these circuits for the maximum noise originating in those circuits. This would all but obscure signals from the antenna that we wish to amplify.

The noise generator can also be used to compare the sensitivity of two different receivers. Most amateur receivers have an S meter which indicates the strength in DB of the received signal. Set the potentiometer (R1) to the point which produces an S7 reading on a receiver's S meter. Without disturbing the adjustment of R1, reconnect the noise generator to the second receiver. If the second receiver's S meter shows a higher reading than S7, the second receiver is more sensitive. If the second receiver's S meter shows a lower reading, this receiver is less sensitive.

Index

A

Absorption as factor in ionization, 47
AC cycle, 38
Air-core inductor, 161
AM, circuitry for, 81
AM radio broadcast band, 40
Amateur building of space hardware, 70
 cost, 70-71
Amateur licenses, 20 (see also "Licenses")
Amateur Satellite Corp., 34, 71 (see also "Space communication")
Amateur satellites, 53 (see also "Space communication")
AMECO Model for Morse Code, 186-187
American Radio Relay League, 19, 60
Amplitude modulation, circuitry for, 81
AMSAT, 34 (see also "Space communication")
 meaning, 251
AMSAT-OSCAR 6, 54 (see also "Space communication")
AMSAT-OSCAR 7, 54-58 (see also "Space communication")
Angstrom, 40-41
Antenna, building your own, 205-218
 80 or 40 meters, choice between, 205
 80 meter doublet, 206-210
  balun, 208-209
  copperweld, 208
  testing, 209-210
  wire size, 208
 satellite signal-squirter, 215-217
 short 40 meter vertical, 210-215
  loading coil, 211-215

Antenna, building your own (cont'd)
 two-meter, 217-218
Antennas, how to understand, 91-111
 collinear and parabolic arrays, 109-110
 current, 91-93
 directivity, 100-103
 distribution of voltage and current, 93-94
 energy, feeding to, 95-97
 helical whip, 104-106
 impedance, 94-95
  eel, electric, as illustration, 95
  phase relationship, 95
  reactance, 94-95 (see also "Reactance ...")
  resistance, 94
 "inverted V," 106-108
 loading coils, 98-99
 masts, 110-111
 mobile, 99
 polarization, 100
 Quad, cubical, 106
 and RF tanks, coupling, 161-165
 theory, 91
 transmatch, 111
 vertical, 97-98
 Yagi, 103-104
ARCOS, 56
Armstrong, Major, 78
Audio chokes, 141-142
Audio oscillator, 252-253
Audion, 84

B

Balun of antenna, 208-209
Band-pass filter, 149-150
Beat-frequency oscillator, 56, 83, 252
Bose, J.C., 75
Building your own set, 195-204
  transmitter, homemade, 195-196
    construction practice, 196-199
    power supply, 202-203
    project, 201-202
    soldering, 200
    triode-pentode, 200-201
    tuning up, 203-204

C

Capacitive reactance, 95
Capacitors, 139-140
Cassette practice tapes for Morse Code, 192
Cat's whisker, 75
Cathode-ray tube in oscilloscope, 248
Ceramic capacitors, 139-140
Chokes, 141-142
Circuitry, guide to, 75-89
  crystal set, TRF receiver from, 75-78
  super-heterodyne receiver, 78-84
    amplitude modulation, 81
    double-conversion, 80
    FM receivers, 80-83
    IF amplifier, 79
    Morse Code and single sideband, receiving, 83-84
  transmitters, 84-89
    FM, 87-88
    single sideband, 88-89
    spark gap, 84-87
Circuits, amateur, detailed guide to, 147-166
  band-pass filter, 149-150
  de-emphasis filters, 148-149
  emphasis filters, 148-149
  filters, types of, 147
  high-pass filters, 147-148
  low-pass filters, 148-149
  mechanical filter, 150
  oscillators, 154-156
  pi-output tanks, 165-166
  power supplies, 150-152

Circuits, amateur, detailed guide to (cont'd)
  RF amplifiers, 158-161
    air-core inductor, 161
    neutralizing capacitor, 161
    slug tuning, 158
  RF tanks and antenna coupling, 161-165
  sideband filter, 149-150
  surge protection ,153
  synthesizers, 156-158
    prescaler, 157
CMOS microprocessor, 68
Collinear array, 109
Colpitts oscillator, 156, 157, 161
Conductors and semiconductors, differences between, 133
Copperweld wire to build antenna, 208
COR of machine, 240
Cost of building satellites, 70-71
Critical frequency, 47-48
Crystal set as origin of TRF receiver, 75-78

D

D region of ionization, 46
De-emphasis filters, 148-149
de Forest, Dr. Lee, 77
Demodulator, 37
Diagrams, guide to, 113-132 (see also "Schematic diagrams.....")
Diodes, 134-135
  Zener, 134-135
Dipole, 101, 103
Dipper, 210-211
Directivity of antenna systems, meaning of, 100-103
Directory, repeater, 240
Discriminator, 82
Disk capacitors, 140
Double-conversion super-heterodyne receiver, 80

E

E layer of ionization, 45-46
Educational aide, OSCAR 6 as, 71
Eel, electric, as illustration of phase principle, 95

INDEX

80 or 40 meters, choice between, 205
80 meter doublet, 206-210
Electrolytic capacitors, 139-140
Electron theory, basic guide to, 115-132
(see also "Schematic diagrams...")
Electron voltmeters, 245-246
Emergencies, history of service in, 19-20, 34
Emphasis filters, 148-149
Energy, feeding to antenna, 95-97
Equipment, amateur test, key factors in, 245-258
  audio oscillator, 252-253
  electron voltmeters, 245-246
  field strength meter, 254-255
  frequency marker, 250-252
  noise generators, 257-258
  oscilloscope, 247-250
  RF probe, 246-247
  SWR meter, 245
  VSWR, measuring, 255-257
    null, 255
Equipment required for space communication, 54-58 (see also "Space communication")

F

F layers of ionization, 46-47
FCC field offices, 21-23, 27-29
FCC Novice Class exam, typical questions on, 167-181
  circuits, 174-176
  electricity, theory of, 173-174
  electronic components, 176-180
  operating procedure, 171-173
  radio phenomena, 170-171
  rules and regulations, 167-170
FET, 138-139
FET-VOM, 246
Field strength meter, 254-255
Filters, types of, 147-150 (see also "Circuits, amateur...")
FM, circuitry for, 80-83
FM transmitters, 87-88
40 meter vertical antenna, short, 210-215
  loading coil, 211-215
Frequency deviation, 88
Frequency marker, 250-252

Frequency Shift Keying, 37
FSK, 37

G

Galena, 75
Galena crystal detector, 19
Galena diode, 134
General Class Exam, preparing for, 219-238 (see also "Technician's Exam ....")
Geosynchronous orbits, 62
Germanium diodes, 134
Grid-dip meter, 210-211
Ground wave, 49

H

Half-wave antenna, building, 206-208
Ham radio, 19-41
  amateur television, 31-32
  emergencies, services in, 19-20, 34
  FCC field offices, 21-23, 27-29
  frequency, 38-39
  frequency and mode allocations, 30
  licenses, 20
    locations for, 21-27
    novice, 20, 30
  meters and megacycles, 40-41
    angstrom, 40-41
    microwave region, 40
    particle radiation, 41
    radar, development of, 40
  novice license, 20, 30
  phone patching, 34
  pioneers, 19-20
  radioteletype, 36-37
    Frequency Shift Keying, 37
    Murray encodement system, 37
  satellites, 34-36
  Slow-Scan television, 32-33
  specialized communication, 31
  wavelength, 39
Hamtronics, 56
Hartley oscillator, 156, 157, 161
Helical whip, 104-106
High frequencies, 50-51
High-pass filter, 147-148

INDEX

I

IF amplifier, 79
Impedance, meaning of, 94-95
   reactance, 94-95 (see also "Reactance ...")
   resistance, 94
Inductive-link coupling, 164
Inductive reactance, 95
Integrated Housekeeping Unit, 68
International Morse Code, studying, 183-193 (see also "Morse Code...")
"Inverted V" antenna, 106-108
Ionization, regions of, 43
   layers, 45-49
Ionosphere, 45

L

Licenses, 20
   locations, 21-27
   novice, 20, 30
Linking experiments, 59
Local oscillator, 79
Low frequencies, 49-50
Low-pass filter, 148-149

M

Machine, 239-242
   COR, 240
Marconi, 75
Masts for antennas, 110-111
"MAYDAY" distress signal, 184
Mechanical filter, 150
Mica capacitors, 140
Microfarads, 142-145
Microhenrys, 142-145
Microwave region, 40
Morse Code, how to benefit from course in, 183-193
   exercises, 191-192
   key, holding, 189-190
      adjusting, 190
   Navy's method, 184-185
   oscillator, 186-187
   practice, rules for, 187-188
   reasons for, 183-184

Morse Code, how to benefit from course in, (cont'd)
   receiving in super-heterodyne circuit, 83-84
   sending, rules for, 188-189
   warming up to key, 191
MOSFET, 138-139, 154, 158
Murray encodement system, 37

N

N-type material, 133 (see also "Semiconductors...")
National Bureau of Standards, 43
Navy Code Study course, how to benefit from, 183-193 (see also "Morse Code...")
Noise generators, 257-258
Novice Class exam, FCC, typical questions on, 167-181 (see also "FCC Novice Class exam...")
Novice license, 20, 30
NPN junction, 136
Null, meaning of, 255

O

Ohm's Law, 139
OSCAR, 20, 34-36, 50, 53-73 (see also "Space communication")
   as educational aide, 71
   locating, 60-62
OSCAR signal-squirter, making, 215-217
OSCARLOCATOR, 60-62
Oscillators, 154-156
Oscilloscope, 247-250

P

P-type material, 133 (see also "Semiconductors...")
Parabolic dish antenna, 109-110
Particle radiation, 41
Phase relationship, 95
Phase III series of satellites, 62-64 (see also "Space communication")
   amateur/professional, 70
   frequencies, best, 66-67

INDEX

Phase III series of satellites, (cont'd)
  geosynchronous orbits, 62
  mechanics, 65-66
  microprocessor, continuing control by, 68-69
    Integrated Housekeeping Unit, 68
  orbits possible, three, 63
  power supply, 67-68
  professional/amateur interaction, 64-65
  technical problem, 64
Phone patching, 34
Pi-network circuit, 165-166
Pi-tuning network, 144
Piezoelectric effect, meaning of, 137
PNP junction, 136
Polarization, meaning of, 100
Prescaler, 157
Project Oscar, 20, 34-36, 50 (see also "Oscar")
Propagation, ham bands and, 43-51
  bulletins, 49
  high frequencies, 50-51
  ionization layers of, 45-49
    absorption, 47
    critical frequency, 47-48
    ground wave, 49
    skip distance, 48-49
  ionosphere, 45
  low frequencies, 49-50
  radiation from sun, 43
    ionization, regions of, 43
    Zurich sunspot number, 43-44
  sunspots, 44-45

Q

Q of tank circuit, 162
Quad, cubical, 106
Quartz crystals, 137
  piezoelectric effect, 137
Questions, typical, on FCC Novice Class exam, 167-181 (see also "FCC Novice Class exam...")
QST, 20
  and amateur television, 31

R

Radar, development of, 40

Radio Amateur's Handbook, 181
Radioteletype, 36-37
RAM, 68
Ratio detector, 82
Reactance, meaning of, 94-95
  capacitive, 95
  inductive, 95
Repeaters, understanding, 239-244
  directory, 240
  finding, 239
  machine, 239-242
    COR, 240
  popularity, increasing, 240
  remote base stations, 242, 244
  rules for responsible operation, 243
Resistance, meaning of, 94
Resistors, 139
RF amplifiers, 158-161
  air-core inductor, 161
  neutralizing capacitor, 161
  slug tuning, 158
RF chokes, 141-142
RF inductors, 142-145
RF probe, 246-247
RTTY, 36-37

S

Satellite signal-squirter, making, 215-217
Satellites, 34-35
Satellites, amateur, 53 (see also "Space communications")
Schematic diagrams, practical guide to, 113-132
  electron theory, 115-132
  quiz, 124
  symbols used, common, 114
Semiconductors and other components, 133-145
  capacitors, 139-140
  chokes, 141-142
  and conductors, differences between, 133
  diodes, 134-135
    Zener, 134-135
  FET, 138-139
  field effect transistors, 138-139
  junctions, 133-134
  MOSFET, 138-139
  N-type and P-type materials, 133

Semiconductors and other components, (cont'd)
  quartz crystals, 137
    piezoelectric effect, 137
  resistors, 139
    Ohm's Law, 139
  RF inductors, 142-145
  temperature, sensitive to, 134
  theory, 133
  transistors, 136-137
    NPN junction, 136
    PNP junction, 136
Semiprofessional technology, development of, 70-71
Sideband filter, 149-150
Silver micas, 140
Sine wave, 38
Single sideband, receiving in super-heterodyne circuit, 83-84
Single sideband transmitters, 88-89
Skip distance, 48-49
Slow-Scan television, 32-33
Slug tuning, 158
Smith, Randall, 60
Soldering to make your own transmitter, 200
Space communication, 53-73
  amateur satellites, history of, 53-54
    AMSAT-OSCAR 6, 54
    AMSAT-OSCAR 7, 54-58
  applications, other, 72-73
  cost, 70-71
  as educational aide, 71
  equipment required, 54-58
    beat-frequency oscillator, 56
    transverter, 56
  experiments, 58-60
    linking, 59
  locating OSCAR, 60-62
  Phase III series, 62-64 (see also "Phase III...")
    amateur/professional, 70
    cooperative effort of amateurs and professionals, 64-65
    frequencies, best, 66-67
    geosynchronous orbits, 62
    mechanics, 65-66
    microprocessor, control by, 68-69 (see also "Phase III....")
    orbits possible, three, 63
    power supply, 67-68
    technical problem, 64

Spark gap transmitter, 84-87
Sputnik I, 53
SSTV, 32-33
Sun, radiation from, 43
Sunspots, 44-45
Super-heterodyne receiver, 78-84
Surge protectors, 153
SWR meter, 245
Symbols, electronic, table of commonly used, 114
Synthesizers, 156-158
  prescaler, 157

T

Talcott Mountain Science Center, 71
Tank circuit Q, 162
Technician's Exam, preparing for, 219-238
  electricity, basic, 227-238
  and general exam, requirements for, 219-220
  propagation and operation, 223-227
  rules and regulations, 220-223
Television, amateur, 31-32
Texas RF Communications, 56
Transistors, 136-137
  NPN junction, 136
  PNP junction, 136
Transmatch, 111
Transmission path, 43-51 (see also "Propagation...")
Transmitter, making your own, 195-196 (see also "Building your own set")
  construction practice, 196-199
Transmitters, 84-89 (see also "Circuitry...")
Transponder frequencies of Phase III series, 66-67
Transverter, 56
TRF, 78-80 (see also "Circuitry...")
2-meter antennas, 217-218

V

Vacuum tube, 84-85, 134
Vertical antennas, 97-98
VHF Engineering, 56
Voltage standing wave ratio, 97
VSWR, 97
  measuring, 255-257

INDEX

W

Wavelength, 39
WIAW, 20
  broadcasts of Morse Code, 192
WWV and WWVH, 49

Y

Yagi, 103-140

Z

Zener diode, 134-135
Zurich sunspot number, 43-44
Zwirko, Rich, 58